CHEMISTRY AND PHYSICS OF CARBON

A SERIES OF ADVANCES

Edited by

Philip L. Walker, Jr. and Peter A. Thrower

DEPARTMENT OF MATERIAL SCIENCES
THE PENNSYLVANIA STATE UNIVERSITY
UNIVERSITY PARK, PENNSYLVANIA

Volume 13

MARCEL DEKKER, INC.   New York and Basel

The Library of Congress Cataloged the
First Issue of This Title as Follows:

Chemistry and physics of carbon, v. 1–
  London, E. Arnold; New York, M. Dekker, 1965– ˙

    v. illus. 24 cm.

  Editor: v. 1– P. L. Walker

  1. Carbon. I. Walker, Philip L., 1928–   ed.

QD181.C1C44   546.681   66–58302

Library of Congress   1
ISBN 0–8247–6359–9

MARCEL DEKKER, INC.

270 Madison Avenue, New York, New York 10016

Current printing (last digit):
10 9 8 7 6 5 4 3 2 1

PRINTED IN THE UNITED STATES OF AMERICA

PREFACE

The editors have worked hard at locating authors to write about diamonds. After all, a series on the "Chemistry and Physics of Carbon"
should be even-handed in its treatment of solids showing tetrahedral
carbon bonding as well as those showing trigonal carbon bonding. It
happens, however, that there has been much more research conducted
on the trigonally bonded carbon structure and, hence, a dispropor-
tionate number of chapters in the series devoted to such solids.

However, perhaps what we have lacked in quantity we have made
up for in quality. In Volume 10 we had an excellent chapter by Bundy
and co-workers on "Methods and Mechanisms of Synthetic Diamond
Growth." In this volume Dr. Gordon Davies has, in our judgment,
produced an epic work on "The Optical Properties of Diamond." What
could be more glamorous and thought-provoking than this subject?
Dr. Davies considers in detail the fundamentals and is able to quantitatively interpret changes in optical properties brought about by
imperfections, nitrogen inclusion, and irradiation.

In the second chapter in this volume we have another valuable
contribution from a worker in the United Kingdom Atomic Energy Agency.
Workers in this agency, over a period of years, have conducted
research on graphite in support of the gas cooled-graphite moderated
nuclear reactor program. It is a classic case of demonstrating the
viability of coupling good research and good engineering to yield
a product useful to mankind. In the use of graphite as a moderator
and a structural member in reactors, it has been necessary to have
a fundamental understanding of its mechanical properties and how
they are affected by fast neutron irradiation. Dr. Brocklehurst,
in his chapter on "Fracture in Polycrystalline Graphite," presents
a definitive account of these subjects.

Peter A. Thrower
Philip L. Walker, Jr.

CONTRIBUTORS TO VOLUME 13

J. E. BROCKLEHURST   United Kingdom Atomic Energy Authority,
Reactor Fuel Element Laboratories, Springfields, Preston,
Salwick, Lancashire, England

GORDON DAVIES   Wheatstone Physics Laboratory, University of
London King's College, London, England

CONTENTS OF VOLUME 13

Preface to Volume 13. iii
Contributors to Volume 13 iv
Contents of Other Volumes vi

THE OPTICAL PROPERTIES OF DIAMOND 1

    Gordon Davies

   I.   Introduction 2
  II.   Chemical Analyses of Diamond 4
 III.   Types of Diamond 6
  IV.   The Optical Properties of Pure Diamond 49
   V.   Simple Vibronic Spectra. 58
  VI.   Nonsimple Vibronic Spectra 74
 VII.   Perturbations of Optical Centers by Nitrogen 92
VIII.   Photochromic Effects 102
  IX.   Donor-acceptor Pair Spectra. 106
   X.   Final Remarks. 109
        Appendix: Absorption, Luminescence, and
        Photoconduction Transitions in Diamonds. 112
        References 128

FRACTURE IN POLYCRYSTALLINE GRAPHITE. 145

    J. E. Brocklehurst

   I.   Introduction 146
  II.   Material 147
 III.   Deformation and Associated Characteristics 150
  IV.   Fracture Under Uniform Stress. 165
   V.   Statistics of Fracture, Size Effects,
        and Nonuniform Stress Conditions 180
  VI.   Effective Work of Fracture, Fracture Toughness,
        and Inherent Defect Size 193
 VII.   Fatigue. 204
VIII.   Notch Sensitivity. 218
  IX.   Effect of Density, Grain Size, and Crystallinity 225
   X.   Effect of Temperature, Atmosphere, and Strain Rate . . . 231
  XI.   Effect of Fast-neutron Irradiation 243
 XII.   Effect of Intercalation. 258
XIII.   Summarizing Discussion 264
        References 272

        Author Index 281
        Subject Index. 293

CONTENTS OF OTHER VOLUMES

VOLUME 1

Dislocations and Stacking Faults in Graphite, S. Amelinckx, P.
        Delavignette, and M. Heerschap
Gaseous Mass Transport within Graphite, G. F. Hewitt
Microscopic Studies of Graphite Oxidation, J. M. Thomas
Reactions of Carbon with Carbon Dioxide and Steam, Sabri Ergun and
        Morris Menster
The Formation of Carbon from Gases, Howard B. Palmer and Charles F.
        Cullis
Oxygen Chemisorption Effects on Graphite Thermoelectric Power,
        P. L. Walker, Jr., L. G. Austin, and J. J. Tietjen

VOLUME 2

Electron Microscopy of Reactivity Changes near Lattice Defects in
        Graphite, G. R. Hennig
Porous Structure and Adsorption Properties of Active Carbons,
        M. M. Dubinin
Radiation Damage in Graphite, W. N. Reynolds
Adsorption from Solution by Graphite Surfaces, A. C. Zettlemoyer
        and K. S. Narayan
Electronic Transport in Pyrolytic Graphite and Boron Alloys of
        Pyrolytic Graphite, Claude A. Klein
Activated Diffusion of Gases in Molecular-Sieve Materials, P. L.
        Walker, Jr., L. G. Austin, and S. P. Nandi

VOLUME 3

Nonbasal Dislocations in Graphite, J. M. Thomas and C. Roscoe
Optical Studies of Carbon, Sabri Ergun
Action of Oxygen and Carbon Dioxide above 100 Millibars on "Pure"
        Carbon, F. M. Lang and P. Magnier
X-Ray Studies of Carbon, Sabri Ergun
Carbon Transport Studies for Helium-Cooled High-Temperature Nuclear
        Reactors, M. R. Everett, D. V. Kinsey, and E. Römberg

VOLUME 4

X-Ray Diffraction Studies on Carbon and Graphite, W. Ruland
Vaporization of Carbon, Howard B. Palmer and Mordecai Shelef
Growth of Graphite Crystals from Solution, S. B. Austerman
Internal Friction Studies on Graphite, T. Tsuzuku and M. H. Saito
The Formation of Some Graphitizing Carbons, J. D. Brooks and
        G. H. Taylor
Catalysis of Carbon Gasification, P. L. Walker, Jr., M. Shelef, and
        R. A. Anderson

VOLUME 5

Deposition, Structure, and Properties of Pyrolytic Carbon, J. C.
        Bokros
The Thermal Conductivity of Graphite, B. T. Kelly
The Study of Defects in Graphite by Transmission Electron Microscopy,
        P. A. Thrower
Intercalation Isotherms on Natural and Pyrolytic Graphite, J. G.
        Hooley

VOLUME 6

Physical Adsorption of Gases and Vapors on Graphitized Carbon Blacks,
        N. N. Avgul and A. V. Kiselev
Graphitization of Soft Carbons, Jacques Maire and Jacques Méring
Surface Complexes on Carbons, B. R. Puri
Effects of Reactor Irradiation on the Dynamic Mechanical Behavior of
        Graphites and Carbons, R. E. Taylor and D. E. Kline

VOLUME 7

The Kinetics and Mechanism of Graphitization, D. B. Fischbach
The Kinetics of Graphitization, A. Pacault
Electronic Properties of Doped Carbons, André Marchand
Positive and Negative Magnetoresistances in Carbons, P. Delhaes
The Chemistry of the Pyrolytic Conversion of Organic Compounds to
        Carbon, E. Fitzer, K. Mueller, and W. Schaefer

VOLUME 8

The Electronic Properties of Graphite, I. L. Spain
Surface Properties of Carbon Fibers, D. W. McKee and V. J. Mimeault
The Behavior of Fission Products Captured in Graphite by Nuclear
        Recoil, Seishi Yajima

VOLUME 9

Carbon Fibers from Rayon Precursors, Roger Bacon
Control of Structure of Carbon for Use in Bioengineering, J. C.
        Bokros, L. D. LaGrange, and F. J. Schoen
Deposition of Pyrolytic Carbon in Porous Solids. W. V. Kotlensky

VOLUME 10

The Thermal Properties of Graphite, B. T. Kelly and R. Taylor
Lamellar Reactions in Graphitizable Carbons, M. C. Robert, M. Oberlin,
        and J. Méring
Methods and Mechanisms of Synthetic Diamond Growth, F. P. Bundy,
        H. M. Strong, and R. H. Wentorf, Jr.

VOLUME 11

Structure and Physical Properties of Carbon Fibers, W. N. Reynolds
Highly Oriented Pyrolytic Graphite, A. W. Moore
Deformation Mechanisms in Carbons, Gwyn M. Jenkins
Evaporated Carbon Films, I. S. McLintock and J. C. Orr

VOLUME 12

Interaction of Potassium and Sodium with Carbons, D. Berger, B.
        Carton, A. Métrot, and A. Hérold
Ortho-/Parahydrogen Conversion and Hydrogen Deuterium Equilibration
        over Carbon Surfaces, Y. Ishikawa, L. G. Austin, D. E. Brown,
        and P. L. Walker, Jr.
Thermoelectric and Thermomagnetic Effects in Graphite, T. Tsuzuku
        and K. Sugihara
Grafting of Macromolecules onto Carbon Blacks, J. B. Donnet,
        E. Papirer, and A. Vidal

THE OPTICAL PROPERTIES
OF DIAMOND

Gordon Davies

Wheatstone Physics Laboratory
University of London King's College
London, England

I. INTRODUCTION 2

II. CHEMICAL ANALYSES OF DIAMOND 4

III. TYPES OF DIAMOND 6
   A. Introduction 6
   B. Type Ia Diamond. 8
   C. Type Ia Diamond: The $\sim$1370-cm^{-1} Peak. 13
   D. Type Ia Diamond: Ultraviolet Spectra. 14
   E. Type Ie Diamond: Atomic Form of Nitrogen. 18
   F. Type Ib Diamond. 24
   G. Type IIb Diamond: Definition and Infrared
      Properties 30
   H. Type IIb Diamond: Ultraviolet Properties. 40
   I. Type IIb Diamond: The Nature of the Accepter. . . . 46
   J. Type IIa Diamond and the Limitations of
      Type Classification 48
   K. Intermediate Type Diamonds 49

IV. THE OPTICAL PROPERTIES OF PURE DIAMOND 49

V. SIMPLE VIBRONIC SPECTRA. 58
   A. General Theory 58
   B. The 3.150-eV Band. 68
   C. The 2.985-eV Band. 71

VI. NONSIMPLE VIBRONIC SPECTRA 74
   A. The 1.673-eV Band. 74
   B. The 1.946-eV Band. 82
   C. The 2.463- and 2.499-eV Bands. 88

VII. PERTURBATIONS OF OPTICAL CENTERS BY NITROGEN 92

VIII. PHOTOCHROMIC EFFECTS 102

IX. DONOR-ACCEPTOR PAIR SPECTRA. 106

X. FINAL REMARKS. 109

   APPENDIX: ABSORPTION, LUMINESCENCE, AND
   PHOTOCONDUCTION TRANSITIONS IN DIAMONDS. 112

   REFERENCES 128

I.   INTRODUCTION

A diamond is synthesized when two larger ones, one male and one fe-
male, come together, in the hills where the gold is. And the diamond
grows larger in the dew of a May morning [1a]. It is indestructible,
unless it is wetted by the warm blood of a billygoat [1b], and it
will make the man who wears it invincible in battle [1a]. If he
loses his way, it will seek out the north for him [1c,d], light his
path [1e], and protect him from fever [1f].

Of course, these are myths, and we can dismiss them as such
without further thought. But other stories sound more plausible.
For example, it is often said that diamonds are individuals, no
two being alike. Since diamond is a natural mineral and likely to
be 2×10^9 yr old [2], it is quite likely that impurities may have
diffused in, radiation damage processes may have been at work, and
each diamond really will be different from all others. If this is
the case, then the study of natural diamonds will be a tedious
exercise in statistics.

Fortunately, it turns out that the vast majority of diamonds
can be classified into a few "types," all diamonds of a given type
having qualitatively the same behavior. To see why things are so
simple, we begin by looking at the impurity content of diamonds.
It is shown that diamonds are really quite pure, with the exception
that nitrogen may be present in up to about 0.3 atomic percent
$(5 \times 10^{26}$ N/m$^3)$ in some specimens. In Sec. III, we describe the
properties of the different types of diamond. Then, knowing the
extrinsic properties of diamond, we are able to deduce the properties
perfect diamond would have (Sec. IV). Diamonds are rich in optical
spectra caused by atomic-sized imperfections. These are discussed
in Secs. V and VI. We see that many of the vibronic properties of
these spectra are now understood in detail, but other spectra, particularly the 1.673-eV (GR1) band, still present challenges.

The nitrogen content in diamond can be high; we have quoted a
figure of 5×10^{26} atoms m^{-3}. The evidence currently available
points to this nitrogen being present, usually, in very small aggregates, perhaps two nitrogen atoms per aggregate. Consequently in
some diamonds, all points lie within ten atomic spacings of a nitrogen aggregate. Weak interactions between the optical centers and

nitrogen can profoundly change the properties of the diamonds. For
example, the luminescence of a diamond can be reduced to very low
levels by energy being transferred from excited optical centers to
the nitrogen. The quenching of the luminescence produces luminescent zero-phonon lines which are sharper than the corresponding
absorption lines, because the linewidths are also produced by
perturbations of the optical centers by the nitrogen. These effects
are discussed in detail in Sec. VII.

Studies of the optical properties of irradiated diamonds can be
greatly confused by photochromic effects. In Sec. VIII, we show
that the origin of these photochromic processes lies in the production of free charges when light is absorbed in the 3.150-eV band
and in the ultraviolet absorption continuum associated with the
1.673-eV center. Finally, in Sec. IX, we briefly discuss the properties of donor-acceptor pairs in diamond.

When describing the optical properties of diamond, we are faced
with two practical difficulties. First, there is the great number
of optical transitions observed in diamond. To combat this, a list
of the optical bands is given in the Appendix, together with brief
descriptions of their properties and references to the appropriate
sections of the text. The Appendix thus serves as an index to the
bands. The second difficulty is that many of the bands are distinguished by a name, such as ND1 or R10. These can be confusing at
first, especially since ND1 and R10 refer to the same band. In this
chapter, we simply label each band by the energy of its zero-phonon
line, followed by the name (if any) in parenthesis. It has to be
remembered that the energies cannot be precisely defined for all
diamonds: One effect of a high nitrogen content is to change the
energy of each zero-phonon line by a small amount. The changes are,
typically, 1 meV to lower energy as the nitrogen concentration increases from zero to 5×10^{26} m^{-3}. In practice, this uncertainty
causes no difficulty when the energies are used in conjunction with
the descriptions given in the Appendix.

On the perennial problem of units, we use electronvolts for the
visible and ultraviolet spectra. In the infrared, we follow the
majority and use cm^{-1} for the absorption spectra of type I diamonds
and millielectronvolts for the properties of the acceptor center in

type IIb diamond. As an apology for using this hybrid system, other
units have been included in the Appendix.

The references are not intended to be comprehensive, but they
should indicate the state of work on the optical properties of diamond in early 1975.

II. CHEMICAL ANALYSES OF DIAMOND

In this section, we briefly review the available data on the impurity
content of diamonds. In the main, the diamonds are of South African
origin, but the major conclusions appear to have universal validity;
for example, workers in the U.S.S.R. report exactly the same optical
properties as are observed in South African specimens.

Table 1 summarizes the results in "inclusion-free" diamonds.
We see that there can be considerable variation in the concentration
of any given impurity, but a fairly clear trend emerges from the different analyses. Nitrogen is the most abundant impurity, followed by
oxygen, then (in some specimens) Si, and next Mg, Fe, Ca, and Al.
Our main problem is to decide exactly what these analyses mean. An
experimenter can only analyze a finite number of diamonds: Some
criteria must be used to select those specimens. He may separate
the more perfect diamonds from those that are seen to contain inclusions under, say, a polarizing microscope at 50x magnification.
But there is no reason, a priori, why inclusions should not occur on
a smaller scale. Generally, the available analyses are of diamond
crystal plus an unknown amount of material, which is contained within
the bulk of the diamond but not actually chemically bonded to the
carbon atoms.

Inclusions in diamond are [17]: garnets (e.g., $Mg_3 Al_2 (SiO_3)_4$),
olivine (i.e., $(Mg, Fe)SiO_3$), enstatite ($MgSiO_3$), diopside ($CaMgSi_2O_6$)
quartz (SiO_2), calcite ($CaCO_3$), kaolinite ($Al_4Si_4O_{10}(OH)_8$), and
ilemite ($FeTiO_3$), and/or hematite (Fe_2O_3). We see that oxygen, silicon, and the same metal ions are recurring as in the list of most
common impurities after nitrogen. It seems reasonable to guess that
these elements are contained in inclusions which are too small to be
seen through simple optical microscopes. Certainly it is these elements whose concentrations are greatly increased in diamonds that

TABLE 1

Synopsis of Chemical Analyses of Inclusion-free Natural Diamond

Element	ppm	Reference	Element	ppb	Reference
N	6–1700	[3]	Cu	8–64	[3]
	50–2500	[5]		0–6000	[4]
	35–2400	[6]		0–2000	[8]
	1500–2800	[9]		0.4–900	[8]
	30–500	[11]			
	47–148	[12]	Mn	0.4–122	[3]
				2–90	[8]
0	30–90	[3]			
			Co	0.5–17	[3]
Si	3	[3]			
	4–5	[4]	V	0.1–8	[3]
	2–108				
			Sc	0.004–2.8	[3]
Mg	0.1–7	[3]	Ni	0.2–1	[3]
	5–9	[4]			
	1–13				
Fe	0.3–8	[3]			
	0–30	[4]			
	0–117				
Ca	0.1–4	[3]			
	0–6	[4]			
	0–3				
Al	0.03–10	[3]			
	5–50	[4]			
	0–50				
	0.1–20	[7]			
Na	0.02–64	[3]			
	0.05–0.13	[8]			
Ti	0.01–0.2	[3]			
	0–5	[4]			
	0–7	[4]			

[a]Data for green Premier Mine diamonds [3] have been omitted since they seem to be anomalously impure.

[b]Qualitative data of Chesley [13], Straumanis and Aka [14], and Bunting and Valkenburg [15] appear in agreement with the table.

[c]Gaseous inclusions have been studied by, e.g., Melton et al. [10].

[d]One analysis [16] reports about 10-ppm gold in one crystal.

definitely contain inclusions. The increase is by 10 to 300 times
over "inclusion-free" diamonds from the same mines [3, 18]. Finally,
there is no optical evidence that O, Si, Mg, Fe, or Ca are present
in the lattice, and any element whose concentration exceeds about
10^{22} m^{-3} (100 ppb) would be expected to be optically active. Some
postulates that Al was optically active are becoming discredited
(Sec. III.G). There is an unconfirmed suggestion by Raal [19] that
Mn causes the very rare 2.3-eV absorption band.

In contrast, the nitrogen is certainly present in the lattice.
This is demonstrated throughout this chapter, especially in the following sections where we treat the properties of nitrogen in detail.

III. TYPES OF DIAMOND

A. *Introduction*

A perfect monovalent crystal cannot absorb radiation with the direct
production of one phonon [20] since there is no first order change in
electric dipole moment produced by any vibrational mode. Absorption
associated with two phonons can occur. Qualitatively, we can think
of one phonon being required to break the translational symmetry of
the crystal, permitting local fluctuations in the charges on each
atom. Radiation can couple to these instantaneous dipoles and so can
be absorbed with the creation of a second phonon. Short wavelength
modes will be most important in these processes. In diamond, this
intrinsic absorption lies predominantly above 1330 cm^{-1}, the peak in
the density of phonon states of perfect diamond [21, 22].

Impurities change all this by destroying the translational symmetry of the crystal and creating local static dipoles. The absorption spectrum may then be as in Fig. 1. The absorption at wavenumbers greater than 1400 cm^{-1} is the intrinsic absorption, while that
below 1400 cm^{-1} is specimen dependent both in strength and shape
(Figs. 1 through 3, 10, and 13). The variation in shape of these
infrared spectra forms the best way of defining the different "types"
of diamond. This type classification is very widely used in diamond
work; it provides a convenient summary of the properties of a diamond.
The limitations in its usefulness in optical work become apparent
in this chapter.

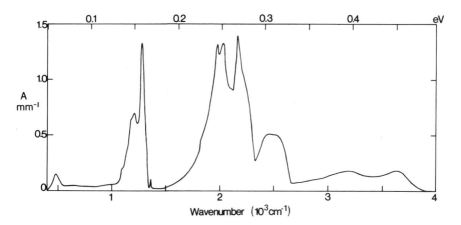

FIG. 1.   Infrared absorption spectrum of a type IaA diamond.
The absorption at wavenumbers greater than 1380 cm^{-1} is present in
all diamonds with the same strength.   The absorption below 1380 cm^{-1}
is specimen dependent.

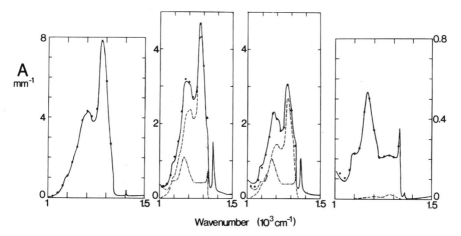

FIG. 2.   A self-consistent fit (...) to the measured type Ia
absorption spectra (—) of seven diamonds is obtained using a two-
component model (N = 2 in Eq. (2)).   The components are shown by the
broken line (- - -) for the A spectrum and the chain line (- · -)
for the B spectrum.   For brevity only four of the seven spectra are
shown.

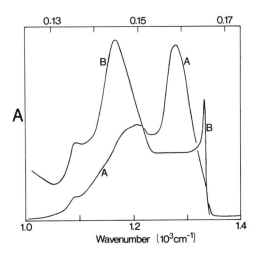

FIG. 3. The absorption induced in the one-phonon range by the A and B nitrogen aggregates, characteristic of types IaA and IaB diamond.

B.   Type Ia Diamond

About 98% of all clear natural diamonds have one phonon infrared spectra like Figs. 1 through 3 [23]. These are type Ia diamonds: A clearer definition follows. The type Ia spectra in Figs. 1 through 3 are obviously superimpositions of more than one component. It is essential to derive the number and form of these components. First, we show that there are only two overlapping bands between 1000 and 1330 cm^{-1}. We can do this without making any assumptions about the shape of these bands.

Let the concentration of defect atoms in the ith defect form be c_i, and let $\mu_i(\nu)$ be the absorption of frequency ν per unit concentration of defect atom in this ith form. For $h\nu < 1330$ cm^{-1} there is negligible intrinsic two-phonon absorption (Fig. 1), so that the total absorption $\mu_T(\nu)$ is

$$\mu_T(\nu) = \sum_{i=1}^{N} c_i \, \mu_i(\nu) \tag{1}$$

The sum is over the N defect species.   Choosing N reference frequencies may solve Eq. (1) for the c_i and resubstitute in the equation,
obtaining

$$\mu_T(\nu) = \sum_{j=1}^{N} a_j(\nu) \ \mu_T(\nu_j) \tag{2}$$

Thus the measured absorption in a particular diamond may be written
as a sum of the absorption coefficient measured in the same diamond
at the N reference frequencies.   The weighting coefficients $a_j(\nu)$
are the same for all type Ia diamonds.   We find [24] that only two
components are required to fit the spectra; with N = 2 and determining the $a_j(\nu)$ by a least-squares fit to the spectra of seven diamonds, we obtain the self-consistent fit shown by the dots in Fig. 2.
The spectra of Fig. 2 have been decomposed into the two components
and the components are compared more clearly in Fig. 3.

We define a diamond with absorption like curve A in
Fig 3 to be a type IaA diamond; and one following
curve B to be a type IaB diamond.

The A or B letter follows the notation of Sutherland et al. [25] who
first observed a qualitative correlation between the peaks in the A
spectrum and between the peaks in the B spectrum.   Note that the
peak at about 1370 cm^{-1} (Figs. 2 and 4) is neither an A nor a B
feature as noted by Clark [26].   Similar results have been given, in
a more qualitative form, by Klyuev et al. [27] and by Sobolev et al.
[28a].

The specimen dependence of the absorption below 1330 cm^{-1} became apparent as early as 1911 when Reinkober [29] found that some
diamonds are transparent at these low frequencies, in contrast to
the opaqueness reported by Ångström (in 1892) [30] and Julius (in
1893) [31].   (The corresponding observation of gross variations in
ultraviolet transmission was also soon made, e.g., by Miller in
1862 [32], Gudden and Pohl in 1920 [33], Levi in 1922 [34], and
Peter in 1923 [35].)   The infrared absorption was studied in more
detail by Robertson et al. in 1934 [36] and in 1936 [37], Ramanathan

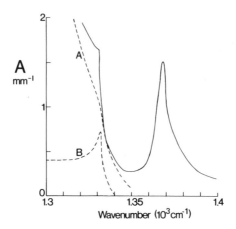

FIG. 4. The $\sim$1370 cm^{-1} peak in a typical type Ia diamond. The
measured absorption spectrum is shown by the solid line. The broken
lines give the A and B components that fit the measured spectrum between 1000 and 1330 cm^{-1}. The $\sim$1370 cm^{-1} line is seen to be asymmetric with an extended wing on its high-energy side.

in 1946 [38], 39] and Sutherland and co-workers in 1945 [40] and
in 1954 [25]. But it was not until 1958 that the chemical origin
of the one-phonon absorption was discovered. Then Kaiser and Bond
[5] correlated the strength of absorption of 1282 cm^{-1} with the total
nitrogen content (Fig. 5). It is important to note that in their
diamonds the A absorption bands were very much stronger than the B
bands (see their Fig. 3). Consequently, their correlation of the
nitrogen concentration with the strength of absorption at 1282 cm^{-1}
refers to the A form:

$$\mu_A (1282 \text{ cm}^{-1}) = (17 \pm 1) \times 10^{-27} \tag{3a}$$

where μ_A is the absorption in mm^{-1} for a unit concentration
(1 atom m^{-1}) of nitrogen in the A form. Alternatively, if N_A percent
of the atoms in a diamond are nitrogen atoms in the A form, then the
absorption in mm^{-1} is

$$A_A (1282 \text{ cm}^{-1}) = (30 \pm 2) N_A \tag{3b}$$

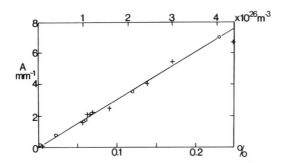

FIG. 5. Correlation between the total absorption at 1282 cm^{-1} and the total measured nitrogen content in type Ia diamonds. The circles are data taken by Kaiser and Bond [5], the crosses by Lightowlers and Dean [6]. The circles are for essentially type IaA diamonds but some of the crosses refer to diamonds with appreciable B nitrogen concentrations. For example, the diamond with 1.5 mm^{-1} total absorption has $A_B(1282 \text{ cm}^{-1}) = 0.3 A_A(1282 \text{ cm}^{-1})$.

In 1964, Lightowlers and Dean [6] confirmed the correlation of Kaiser and Bond between the absorption at 1282 cm^{-1} and the total nitrogen content, using nine diamonds. Some of their diamonds had strong B bands (e.g., D2, see their Fig. 2), but these diamonds still gave the same relationship between the absorption at 1282 cm^{-1} and the total nitrogen concentration. This result implies that the B bands are also caused by nitrogen [24]. Taking account of experimental uncertainties, this B form of nitrogen has an absorption at 1282 cm^{-1} per unit concentration, which is related to $\mu_A(1282 \text{ cm}^{-1})$ by [24]:

$$0.6 < \frac{\mu_B(1282 \text{ cm}^{-1})}{\mu_A(1282 \text{ cm}^{-1})} < 1.6 \tag{4a}$$

Defining A_B as for A_A in Eq. (3b), we have

$$A_B(1282 \text{ cm}^{-1}) = (33 \pm 15) N_B \tag{4b}$$

Sobolev and Lisoyvan [41] confirmed these results by directly measuring the total nitrogen content in diamonds whose infrared spectra showed only the B bands. They found

$$A_B(1175 \text{ cm}^{-1}) = 23 N_B$$

or   (5)

$$A_B(1282 \text{ cm}^{-1}) = 8.5 N_B$$

Although the factor is different in Eqs. (4b) and (5), we can
at least estimate the concentrations of nitrogen in the A and B
forms by decomposing the infrared spectra into the A and B components and then applying Eqs. (3) and (4b) or (5). Note that we can
measure the absorption spectra at room temperature, cooling to 5 K
gives no measurable change in intensity of the A bands and moves them
by only about 1 to 2 cm^{-1} to higher energy [5]. The temperature in-
dependence of the A bands is also reported by Charette [4]. The
B bands also appear to be largely unaffected by temperature changes,
but the 1332 cm^{-1} line is sufficiently sharp for a small increase in
wavenumber of 2 cm^{-1} to be detectable on cooling from 500 to 80 K [4].

Sometimes, for example when dealing with the annealing of radiation damage, it is important to know the relative concentrations of
A and B nitrogen in a type Ia diamond. The ratio can easily be
found using the data of Fig. 6. Both the A and B bands have absorption peaks near 1190 cm^{-1} so that for all A to B ratios there will
be a point of maximum absorption somewhere near 1190 cm^{-1}, say at
wavenumber x. The strongest A feature, at 1282 cm^{-1}, coincides with
a flat absorption region in the B spectrum, so that a peak will
always be present at 1282 cm^{-1} as long as A nitrogen is present.
The ratio of total absorption measured at x and at 1282 cm^{-1}
uniquely determines the ratio of A to B nitrogen concentrations.
In Fig. 6, these concentrations have been expressed in terms of the
absorption at 1282 cm^{-1}, as in Eqs. (3b), (4b), and (5).

The major conclusions in this section are that the nitrogen in
type Ia diamond is generally present in two forms and that these
forms are identifiable from their infrared spectra. It should be
emphasized that these are among the best established data in diamond
science.

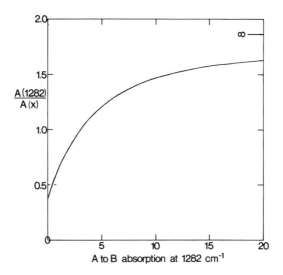

FIG. 6. Data for the decomposition of type Ia spectra into the
A and B components. The total measured absorption at 1282 and
near 1180 cm^{-1} A(1282)/A(x) is compared with the strength of absorption in the A and B components at 1282 cm^{-1}. The limit for infinite
A to B absorption is shown by ∞.

$$C. \quad Type\ Ia\ Diamond:\quad The\ \sim1370\text{-}cm^{-1}\ Peak$$

In addition to the A and B bands, type Ia diamonds very often show

an absorption peak at about 1370 cm^{-1} (Figs. 2 and 4). The precise

wavenumber $\bar{\nu}$ at the peak varies with nitrogen content, following

$$\bar{\nu} = 1359 + 2.8\ A(1282\ cm^{-1}) \tag{6}$$

where A is the absorption in mm^{-1} at 1282 cm^{-1}. (Sobolev et al. [42]

have suggested $\bar{\nu}$ increases as the size of the platelets, discussed

in Sec. III.E, decreases. Their evidence is rather thin, but taken

with Eq. (6), it implies that the mean platelet diameter decreases

with increasing nitrogen content. The few data available [43, 44]

are not fully consistent with this, although there may be a trend

to smaller platelets with higher nitrogen concentrations.)

It seems to be a rule that diamonds never show the ~1370 cm^{-1}

peak unless there is A and/or B infrared absorption as well. On

the other hand, diamonds with strong A and/or B absorption do not
necessarily have a $\sim$1370 cm^{-1} line, although there is a tendency
for the line to be stronger in diamonds with stronger B bands.

By studying the infrared spectra of those type Ia diamonds that
have been chemically analyzed [5, 6], we find [45] that there is no
evidence that the $\sim$1370 cm^{-1} line is produced by nitrogen impurity.
Taking experimental uncertainties into account, we can say that a
diamond with

$$\frac{A(1282 \text{ cm}^{-1})}{A(\sim 1370 \text{ cm}^{-1})} > 1.3$$

has less than 10% of its nitrogen in the form that gives the
$\sim$1370-cm^{-1} peak.

The shift in peak wavenumber with nitrogen content [Eq. (6)]
suggests that the $\sim$1370-cm^{-1} line is sensitive to crystal strains,
so that it can couple to the strain field of the nitrogen (cf.
Sec. VII). We, therefore, expect the line to be temperature dependent. (For example, if it is due to a localized vibration, Eq. (6)
indicates anharmonicity in the vibrational potential.) Charette [4]
reports a movement of the peak to lower wavenumber with increasing
temperature. The shift from 80 to 500 K is about −4 cm^{-1}, the shift
rate increasing as the temperature increases (much as zero-phonon
lines behave, e.g., Fig. 26). The integrated intensity of the line
and also its peak height decrease reversibly by about 30% over the
same temperature range [4]. Similar, but more qualitative data have
been given by Sutherland et al. [25]. No permanent change in
strength results from heating to 2000 K [25] but heating to 2530 K
under a stabilizing pressure of 48 kbar reduces the line by 80% [46].
There do not seem to be lattice-phonon sidebands associated with the
$\sim$1370 cm^{-1} line, although these would be difficult to detect because
of the intrinsic two-phonon absorption.

D. Type Ia Diamond: Ultraviolet Spectra

A perfect diamond would not absorb visible and ultraviolet radiation
as long as hν < 5.5 eV, the band gap energy. The absorption spectrum
would appear as in the bottom curve of Fig. 7. The nitrogen impurity

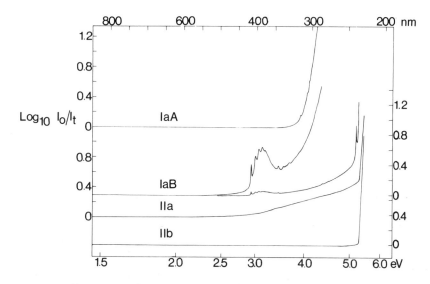

FIG. 7. Typical survey spectra of the visible and ultraviolet room temperature absorption of type IaA, IaB, IIa, and IIb diamonds, taken on a Perkin-Elmer 402 Ultraviolet-Visible Spectrophotometer. The type IaB diamond is shown before and after thinning by a factor of ten. A type IIb diamond may show decreasing absorption from the red to the blue spectral ranges (see the analogous photoconduction spectra, Fig. 15). For type Ib absorption see Fig. 11.

changes this; a pure type IaA diamond absorbs as in the top curve of Fig. 7. The "secondary absorption edge" of type IaA diamond begins near 3.7 eV at room temperature and rises rapidly so that [5, 47]

$$A_A(4.04 \text{ eV}) = 0.5 \ A_A(1282 \text{ cm}^{-1}) \tag{7a}$$

and

$$A_A(4.76 \text{ eV}) = (8.4 \pm 1.5) \ A_A(1282 \text{ cm}^{-1}) \tag{7b}$$

Like all secondary absorption edges in diamond, the type IaA edge is structured. The peaks at 4.05 and 4.19 eV correlate in strength with the A nitrogen concentration [5, 25] and so may provide a useful means of studying the structure of the A nitrogen center. Unfortunately, they have not yet been studied in detail. At low temperature and high resolution the structure is seen to have the linewidths typical of transitions between the electronic states

localized at defects in diamond, for example the 3.765-eV line has
a full width at half height of 3 meV in Fig. 8. These lines mask
the precise threshold of the type IaA secondary absorption edge;
a linear extrapolation of the high energy part of the edge suggests
a threshold near 4 eV at 80 K. Clark et al. [48] estimated the
threshold to be at 3.74±0.05 eV, but their spectra were obtained at
lower resolution and much of the structure was blurred out.

Since nitrogen in diamond has an excess of one valence electron
per atom, we would expect the type IaA secondary edge to be caused
by nitrogen to conduction band transitions. These could be observed
by photoconduction experiments, but we must remember that there is
no simple quantitative relation between photoconduction and absorption, since quantities like the lifetime of the free carriers enter
into the photoconduction. Many type IaA diamonds give a measurable
photoconduction at photon energies less than 3.8 eV, in a region of
essentially zero absorption [33, 36, 50, 51]. There is a pronounced
decrease in this photoconduction at about 3.7 eV (at room temperature)

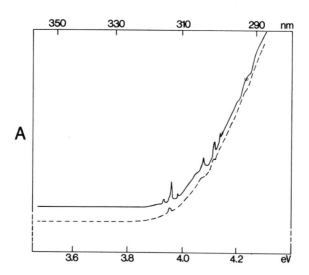

FIG. 8. High-resolution spectrum of the secondary absorption
edge of a type IaA diamond at 80 K (solid line) and 300 K (broken
line). The absorption is essentially zero at the extreme left of
each spectrum.

coinciding with the onset of the internal transitions at the nitrogen (Fig. 8). At higher photon energies, $h\nu \gtrsim 4.05$ eV, there is a
rapid rise in photoconduction [33, 36, 49-52] as we enter the true
secondary absorption continuum. The free carriers have been identified [49, 53-55] as electrons, as expected.

This interpretation of the photoconduction data seems simple
enough but disagrees fundamentally with that put forward by Denham
et al. [49]. They believed that the photoconduction in a type Ia
diamond was solely due to the small concentration of paramagnetic
nitrogen present in most type Ia diamonds. We leave a more detailed
discussion of their paper to Sec. III.F where this paramagnetic
nitrogen is discussed in detail. Here we note only that their argument was based largely on the photoconduction threshold in type Ia
diamond occurring at 4.05 eV in contrast to the absorption threshold
which they give as 3.7 eV. As we have seen, the true secondary absorption edge is probably at a much higher energy.

To summarize, A aggregates of nitrogen produce a secondary absorption edge which starts at about 4 eV and is still increasing [56]
at the fundamental edge at 5.5 eV. The threshold is masked by internal transitions at the nitrogen. The secondary absorption edge is
believed (here) to give the photoconduction, but this requires confirmation by high-resolution, low-temperature photoconduction measurements.

It is more difficult to establish relationships like Eq. (7)
for type IaB diamonds. As the B nitrogen increases, the 2.59-eV(N2)
and 2.985-eV(N3) bands also usually increase, although there is not
a strict relationship between them [47]. The 5.26-eV(N9) bands grow
with a strength closely proportional to the B band strength [47].
All these bands obscure the shape of the B ultraviolet continuum.
It appears, though that [47]

$$A_B(4.76 \text{ eV}) < A_B(1282 \text{ cm}^{-1}) \tag{8}$$

This is illustrated by Fig. 9 and agrees with the generally accepted
qualitative view that type IaB diamonds are much more transparent in

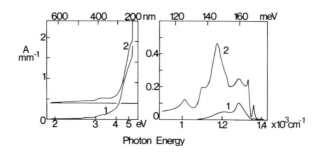

FIG. 9. Ultraviolet absorption spectra of two diamonds (at
left) compared with their infrared absorption (at right). Type IaA
diamonds like specimen number 1 give rise to a strong continuum
in the ultraviolet. A specimen like number 2 with the same strength
of A infrared absorption but much larger B absorption produces
essentially the same strength of continuous absorption. (Superimposed on the continuum of a type IaB diamond is, usually, a 2.985-eV
band, Fig. 20, and, apparently always, a 5.26-eV band, Fig. 21.)

the ultraviolet than type IaA specimens [41]. One practical difficulty with measuring the absorption associated with the nitrogen
near, say, 5 eV is that specimens of low-nitrogen contents are
required. Specimens of this kind often have their own continuous
absorption, as shown by the curve labeled IIa on Fig. 8.

E. Type Ia Diamond: Atomic Form of Nitrogen

So far we have seen that the nitrogen in type Ia diamond is generally
present in two predominant forms, which we have labeled A and B and
can identify from their infrared spectra. It still remains to identify the atomic structure of these A and B aggregates. The shapes
of the A and B bands contain this information, in principle, but
they are both formed by the A and B nitrogen resonating in the lattice;
there are no sharp local mode peaks that would be more characteristic
of the defects. For example, the very sharp B peak at 1332 cm^{-1}
(Fig. 3) is only a resonance at the Raman frequency of diamond. In
the ultraviolet spectrum, the lines at the edge of the A continuum
near 4 eV (Fig. 8) have not yet been investigated. There are no
no sharp visible or ultraviolet lines associated with the B bands;
it seems unlikely that the 5.26-eV(N9) bands are associated with

the B bands, even though they are proportional in strength [47].
Consequently at the moment, there are no direct optical data on the
structure of the A and B forms. But other evidence provides limits
to the possible structures.

First, we know that the A and B forms are both nonparamagnetic.
Since we can exclude the possibility that the nitrogen atoms are
charge-compensated, both the A and B forms of nitrogen must contain
even numbers of nitrogen atoms to allow spin pairing.

The lattice constant of diamond, a, is a known function of the
nitrogen content for type IaA diamond [5]:

$$a = 0.356683 + 5.13 \times 10^{-6} \, A_A (1282 \text{ cm}^{-1}) \text{ nm} \tag{9}$$

where A_A is in mm^{-1}. This expansion is such that if the nitrogen
substitutes for carbon atoms, the increase in mass of the diamond is
balanced by its increase in volume, and an almost constant density
results. This constancy of density is observed for gem quality
type IaA diamonds [5, 57] so that the A form of nitrogen is a largely
substitutional form. A similar density study has not been made for
type IaB diamond, although Sobolev and Lisoyvan [41] could not as-
cribe any lattice constant change to the B form of nitrogen. (We
can note that the use of gem quality diamonds in density experiments
is essential if meaningful results are to be obtained [58, 59].)

The evidence is thus that the A form, at least, is a largely
substitutional aggregate containing an even number of nitrogen atoms.
The main problem is the size of the aggregate. The optical evidence
is that the A aggregates are very small. This is based first on the
way the nitrogen quenches luminescence in type IaA diamond (Sec. VII).
Second, the trapping of radiation damage at the A aggregates during
annealing produces the invariant 2.463-eV(H3) absorption band, suggesting there are very few inequivalent points around the nitrogen
at which the radiation damage can be trapped.

Some negative evidence for small aggregates comes from the
high-resolution microscopy of Evans [60]. He could not detect the
strain fields of small aggregates of nitrogen in a type Ia diamond,

implying that if they exist, the aggregates are below the limiting
detection size, which Evans estimated at about 8 nitrogen atoms per
aggregate.

What electron micrographs do show [43] is the presence in many
type Ia diamonds of large "platelets," condensations of atoms lying
on 100 planes. Each platelet can be up to 100 nm in diameter and
displaces the lattice on either side of the platelet, the total displacement being a/3, where a = 0.3567 nm is the length of the unit
cell [61, 62]. The presence of platelets was predicted by Frank [64]
in 1956 to explain anomalous spikes often seen in the x-ray diffraction photographs of type Ia diamonds [61, 65-69]. The difficulty
with investigating the platelets is that electron microscopy requires
specimen thicknesses of 200 nm, which is far too thin for most other
useful techniques. Consequently, comparison of electron microscope
and other data can be misleading because of specimen inhomogeneity.
However, Evans and Wright [44, 63, 70] have obtained a reasonably
linear relationship between integrated x-ray spike intensity and the
total platelet area per unit volume. This confirms that the platelets produce the x-ray spikes. (Note that small platelets with diameters of a few nanometers will produce x-ray spikes which are too
broad to be observed [44, 71].) Sobolev et al. [28a] and Rainey [28b]
have correlated the spike intensity with the absorption at the peak
of the $\sim$1370-cm^{-1} line. These two correlations, of Evans and Wright,
Sobolev et al., and Rainey, imply that that the platelets give rise
to the $\sim$1370-cm^{-1} line, and so, from Sec. III.C, the platelets do not
contain a significant fraction of the nitrogen in a diamond.

Very often one sees in the literature references to "platelet
nitrogen." This term refers to the assumption that the nitrogen is
present predominantly in the form of platelets, with the remainder,
a very small fraction, being in the single atomic form (Sec. III.F).
As we have seen, the available evidence shows this assumption to be
incorrect. The nitrogen platelet idea arose from the typical concentrations of nitrogen in a type Ia diamond, around 0.1 atomic %,
being similar to the concentration of impurity required to give the
x-ray spikes [61] and, equivalently, being similar to the concentration

of atoms measured in the platelets seen in electron micrographs [43].
In the early 1960s, models of platelets were constructed on the basis
of all the nitrogen being in the platelets [72-74], and the nitrogen
platelet idea became firmly entrenched until Sobolev et al. [28]
actually made quantitative measurements in the late 1960s. But we
can note that even before this there was "strong evidence that the
same cause operates in producing the extra absorption found in the
infrared and ultraviolet spectra of all type I diamonds, but that
additional factors may be the cause of the extra streaks . . . found
in the x-ray photographs of certain diamonds" [25].

It was a gross oversimplification to try to understand all the
properties of type Ia diamond on the basis of just one defect [75].
With the advantage of hindsight, we can see that the concept of all
the nitrogen being in the platelets should never have been accepted;
certainly it should now be forgotten.

In saying that the majority of the nitrogen is not in the platelets, we are not excluding the possibility that they contain a small
fraction of it, but this must be less than 10% of the total nitrogen
in a type Ia diamond (Sec. II.C). So far no one has observed platelets (or the $\sim$1370-cm^{-1} peak) in the absence of the A or B bands, so
that nitrogen most likely plays some role in the formation of the
platelets. Heating to high temperatures under a *hydrostatic* pressure
(e.g., 2530 K, 48 kbar) has no effect on the x-ray spike intensity
(or, equivalently, on the platelet area per unit volume) [44, 46, 76].
Nor does the heating affect the A or B bands [46]. However, the
$\sim$1370-cm^{-1} peak is substantially reduced [46]. Simultaneously, absorption bands with the appearance of the 1.945- and 2.463-eV (H3)
bands are created [46]. Both these bands involve nitrogen. A large
increase in the paramagnetic nitrogen concentration may also be observed [77]. A tentative interpretation is that it is the small concentration of nitrogen trapped in the platelets, which gives the
$\sim$1370-cm^{-1} peak, and that this nitrogen is liberated by the high-
temperature treatment. The correlation of platelet area per unit
volume with the $\sim$1370-cm^{-1} peak (Sec. III.C) must only apply to diamonds not subjected to these high temperatures and pressures.

Some experimenters do report decreases in x-ray spike intensities following heat treatment [78]. This effect is caused by relief of the strain fields near the platelets following plastic deformation, and so occurs when shear stresses are present in the high-pressure cell [44]. The yellow or green color of diamonds after high-temperature annealing [76, 77] is largely caused by Rayleigh scattering in addition to the bands near 1.945 and 2.463 eV [46].

Returning to the question of the size of the A and B nitrogen aggregates, it seems at the moment that the most promising probe into their structure is by thermal conductivity measurements. Turk and Klemens have recently argued that their conductivity data show the nitrogen is predominantly in small aggregate form [79]. Although this conclusion is acceptable here, their derivation of it is not.

The thermal conductivity of a gas, according to elementary kinetic gas theory is $K = (1/3)C \, v^2\tau$, where C is the specific heat per unit volume, v is the mean molecular speed, and $v\tau$ is their mean free path. Applying the same ideas to the phonon gas of a dielectric crystal, the conductivity becomes

$$K = \frac{1}{3} \int d\omega \, C(\omega) \, [v(\omega)]^2 \, \tau(\omega) \tag{10}$$

$C(\omega) \, d\omega$ is the contribution to the specific heat per unit volume from phonons of angular frequency ω. The integral runs over all phonon frequencies. Equation (10) is usually evaluated using a Debye spectrum of phonons. The speed of all phonons is then the same, $v(\omega) = v$, and there are $3\omega^2 \, d\omega/2\pi^2 v^3$ transverse and longitudinal acoustic modes per unit volume. Then

$$K = \frac{\hbar^2}{2\pi^2 vkT^2} \int_0^{\omega_D} \frac{d\omega \, \tau(\omega)\omega^4 \, \exp(\hbar\omega/kT)}{[\exp(\hbar\omega/kT) - 1]^2} \tag{11}$$

The relaxation time $\tau(\omega)$ is made up of several terms representing scattering of phonons at the crystal boundary, phonon-phonon scattering, and scattering at defects in the crystal. We are

particularly interested in scattering from small groups of substitutional atoms as the A form of nitrogen appears to be. These give
a relaxation time [80]

$$\tau_S(\omega) = \frac{\pi \, v^3}{\text{fn} \, a^3} \left(\frac{\Delta M}{2M} + \gamma e\right)^{-2} \omega^{-4}$$ (12)

where

> $\gamma = 0.9 =$ Grüneisen parameter [81]
> $a = 0.3567$ nm = unit cell length
> $f =$ fraction of atoms in the crystal of the defect species
> $n =$ number of defect atoms per aggregate

The defect atom's mass is $(M + \Delta M)$, M being the ^{12}C (host lattice)
mass. The fractional volume change per defect atoms is e.

By fitting the measured curve of K against T by Eq. (11), we
can determine the unknowns in Eq. (12), since a large part of the
thermal resistance in the range 50 to 200 K derives from this substitutional atom scattering. However, the fit will only be meaningful
if the Debye approximation is valid. In diamond, the transverse
acoustic modes have far lower energies than the longitudinal acoustic
modes; the ratio is nearly 1:2 at the $< 1\ 1\ 1 >$ zone boundary [21].
Consequently, the use of a single density of states function
$3\omega^2 \, d\omega/2\pi^2 v^3$ is an oversimplification. By temperatures of 200 K,
there are important contributions to the integrand in Eq. (11)
from phonons of $\hbar\omega \sim 100$ meV, and these are by no means Debye phonons
[21, 22]. As in Sec. V.A, integrals over phonon states seem to fit
measured data without necessarily meaning anything.

Turk and Klemens [79] fitted the data of Slack [82] using aggregates of nitrogen arbitrarily set to contain eight nitrogen atoms
each ($n = 8$). They introduced platelet scattering but treated the
plates as being entirely composed of nitrogen (their Eq. (31)). The
total nitrogen concentration was treated as a variable and gave a
fit with a nitrogen concentration only two-thirds that measured by
Slack. Finally, their derivation of six nitrogen atoms per cluster
using Agrawal and Verma's fit [83] to the data of Berman et al. [84]
is based on a nitrogen concentration obtained from an electron

microscope measurement of the platelet areas [85], which is irrele-
vant.

The thermal conductivity technique seems to have the power to
determine the size of the nitrogen aggregates, but what is required
is work on specimens, such as that shown in Fig. 1, which contain
a known concentration of one form of nitrogen and do not have the
complication of platelets. Finally, we can note that ^{13}C isotope
scattering will be significant at nitrogen concentrations less than
about 0.1 atomic %, if n is small as would seem likely.

To summarize, we have seen in this section that there are no
established models for the A and B forms of nitrogen. But, since
almost pure type IaA diamonds are readily found (e.g., Fig. 1), it
seems reasonable to guess that the A form is the simplest possible
structure, which is then a pair of substitutional nitrogen atoms,
to allow spin pairing. The B form must then be some larger aggregate, since it too is composed of an even number of nitrogen atoms.
This appears to be consistent with the usual occurrence of other
aggregates of nitrogen, such as the 2.985-eV(N3) centers, in diamonds
containing B aggregates. The B form may be, say, four substitutional
nitrogen atoms. We should emphasize that these structures are tentative, contrary to the impression given in many papers from the
U.S.S.R., that the A form definitely is a pair of nitrogen atoms.
We have dismissed the platelets as being the major depository of the
nitrogen.

F. Type Ib Diamond

*We define a type Ib diamond as one whose infrared
spectrum is as Fig. 10.*

Only about 0.1% of natural diamond is of this type [23], but the
great majority of nitrogen bearing synthetic diamond are type Ib
[86-90]. Diamond coat [91], a brown, fibrous [92] outer shell on
many natural diamonds is a mixture of types Ia and Ib diamond
[12, 93, 94], with other impurities present in optically significant
concentrations [95].

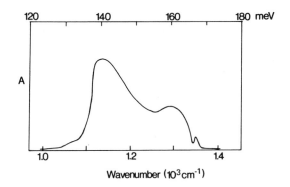

FIG. 10. Type Ib infrared absorption. Data by Dyer et al. [23].

The intensity of the type Ib infrared band correlates with the strength of the electron paramagnetic resonance signal due to single substitutional nitrogen atoms [23]. Chrenko et al. [89] give:

$$A_{Ib}(1130 \text{ cm}^{-1}) = 44N_{Ib} \tag{13a}$$

while Sobolev et al. [88] find

$$A_{Ib}(1130 \text{ cm}^{-1}) = 230N_{Ib} \tag{13b}$$

where N_{Ib} is the percentage of atoms in the diamond that are type Ib nitrogen, and A_{Ib} is the absorption in mm^{-1}. Sobolev et al. used small synthetic diamonds of 0.1 to 1 mg, while Chrenko et al. used far larger synthetic specimens of 14 to 130 mg each. The result of Chrenko et al. would therefore be expected to be more accurate.

The nitrogen also produces absorption in the visible and ultra-violet (Fig. 11). This continuum begins near 1.7 eV, rises rapidly at 4 eV, and merges with the fundamental absorption edge at 5.5 eV. The rise at 4 eV shows much structure, which has not yet been fully investigated (Fig. 12). In a given type Ib diamond, the absorption of the continuum is related to the infrared absorption by:

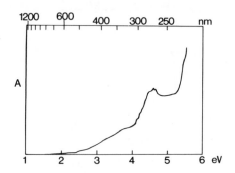

FIG. 11. Type Ib visible and ultraviolet absorption. Data by
Dyer et al. [23]. See also Fig. 12

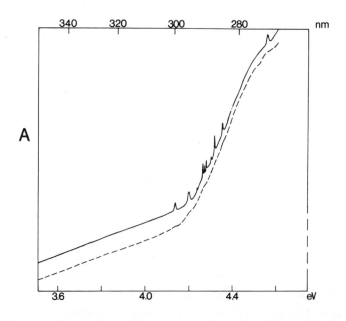

FIG. 12. Type Ib ultraviolet absorption at the sharp rise near
4 eV at high resolution. Solid line at 80 K, broken line at 300 K.
The zero of absorption is not defined in the diagram (see Fig. 11).

Chrenko et al. [89]

$$A_{Ib}(4.6 \text{ eV}) = 45 \ A_{Ib}(1130 \text{ cm}^{-1}) \tag{14a}$$

Sobolev et al. [88]

$$A_{Ib}(2.64 \text{ eV}) = 1.4 \ A_{Ib}(1130 \text{ cm}^{-1}) \tag{14b}$$

or

$$A_{Ib}(4.6 \text{ eV}) \sim 21 \ A_{Ib}(1130 \text{ cm}^{-1}) \tag{14c}$$

The step from Eq. (14b) to (14c) has used [23]
$A_{Ib}(2.64 \text{ eV}) \sim 0.7A_{Ib}(4.6 \text{ eV})$. According to spectra given by Dyer
et al. [23], the infrared and ultraviolet absorptions are related by
$A_{Ib}(4.6 \text{ eV}) \sim 1.5A_{Ib}(1130 \text{ cm}^{-1})$. There is presumably a misprint on
one of their diagrams.

Most of our knowledge of the nitrogen comes from the epr [96-98]
and ENDOR [99] work. The resonance derives from the electron excess
of the nitrogen. Its spin, s = 0.5, couples with the nuclear spin
I = 1 of ^{14}N to give a triplet of lines in the simplest case (magnetic field along < 1 0 0 >). Weak structure in the epr is observed
from interactions involving the I = 0.5 spin of the carbon isotope
^{13}C, which is about 1% abundant in diamond, and also interactions
involving the 0.365% abundant ^{15}N of spin I = 0.5. Nuclear spins of
I = 1 are rare, and this combined with chemical analyses [89] iden-
tifies the impurity beyond reasonable doubt.

Rotation of the magnetic field away from a < 1 0 0 > axis creates
further epr lines as an orientational degeneracy is lifted [96].
The degeneracy comes from one C-N bond being selected as a hyperfine
axis through a Jahn-Teller effect [100, 101]. If the nitrogen atom
was constrained to the tetrahedral substitutional site, its excess
valence electron would enter a triply degenerate T_2 orbital. Linear
coupling to E or T_2 vibrational modes lifts the degeneracy by lowering the symmetry, and a more stable state of lower total energy can
be achieved. A trigonal distortion results when the coupling to T_2
modes is stronger than coupling to E modes [102]. For strong linear

coupling, when the Jahn-Teller energy $E_{JT} \gtrsim 6\hbar\omega$, the ground T_2 vibronic state is joined by an A_1 vibronic state [103], the combined four-fold degeneracy corresponding to the static distortion along one of the four C-N axes. The epr data thus indicate $E_{JT} \gtrsim 6\hbar\omega$, or, taking the dominant phonons to be $\hbar\omega \sim 165$ meV, $E_{JT} \gtrsim 1$ eV. Photoionization of the nitrogen removes the Jahn-Teller-active electron so that E_{JT} contributes to its binding energy. The photoconduction threshold will therefore be at $h\nu \gtrsim 1$ eV rather than ~ 0.4 eV as expected from effective mass theory (see Sec. III.G). Farrer [104] has found photoconduction due to electrons starting at 1.7 eV, at the same energy as the type Ib secondary absorption edge (Fig. 11). The origin of the rapid increase in absorption at 4 eV is not known but could lie in valence band to nitrogen transitions.

Between the 4 stable < 1 1 1 > distortions there will be six barriers along the < 1 1 0 > axes. Öpik and Pryce [102] calculate the barrier to be $E_B = 0.25E_{JT}$ for purely T_2 mode coupling. Taking $E_{JT} \sim 1.5$ eV, $E_B \sim 0.4$ eV. The value measured from thermal reorientation of the epr hyperfine axis is $E_B \sim 0.7$ eV [105, 106]. This is acceptably close to the theoretical value as two conditions implicit in the theory are probably violated. The Jahn-Teller energy is sufficiently large that mixing of different electronic states will occur, in contrast to the theoretical treatment of an isolated T_2 triplet. Nonlinear coupling will also become important at the large displacements of the nitrogen, estimated to be +10% [96, 107] to +26% [101] of the C-C bond length. (The plus sign (+) indicates that the chosen C-N bond is lengthened by the distortion.)

A large Jahn-Teller coupling is predicted from the extended Huckel theory calculations of Watkins and Messmer [101]. These calculations should only be treated as a qualitative guide to reality [108]. The semiquantitative nature of their results is shown by the energy gap of their crystal coming out as 9.5 eV instead of 5.5 eV. Using a diamond composed of 34 carbon atoms and one central nitrogen atom, they obtained $E_{JT} \sim 5$ eV and the barrier energy $E_B \sim 1$ eV. The possibility of such a large Jahn-Teller effect derives from the

strongly localized nature of the donor electron orbital, the electron
being primarily localized on the < 1 1 1 > carbon neighbor [101, 109].

Related to the Jahn-Teller distortion is the form of the vibrational spectrum of the nitrogen atom. Referring to Fig. 10, we see
that a localized mode is usually observed at 1350 cm^{-1} in type Ib
diamonds. The strength of this peak is not proportional to the
strength of the Ib absorption showing that in fact they have different origins. This is confirmed by the energy of the 1350 cm^{-1} mode
being unchanged by ^{15}N doping instead of the more common isotope
^{14}N [110].

We must now return to the suggestion by Denham et al. [49]
(Sec. III.D) and by Dean [111] that the type Ib donor binding energy
is ∿4 eV against the value of 1.7 eV used previously. Denham et al.
reported a difference in energy between the thresholds of the type Ia
secondary absorption edge (3.7 eV) and the photoconduction threshold
(4.05 eV) in type Ia diamonds. This led them to conclude that the
paramagnetic nitrogen present in most type Ia diamonds was responsible
for the photoconduction and that their ionization energy was 4.05 eV.
We have seen (Sec. III.D) that the type IaA edge is not at 3.7 eV
but is closer to 4.05 eV, removing their problem. But even within
their scheme, it is not clear why the absorption peaks of the type Ib
edge near 4 eV should become thresholds in the photoconduction spectrum at the resolution they claim (their Fig. 6). Denham et al.
reported that the photoconduction of a type Ib diamond was weak but
also had a threshold at 4.05 eV. Examination of their type Ib specimen shows that in fact most of its nitrogen is in the A form and a
threshold at 4.05 eV would be expected. Finally, acceptance of the
photoconduction interpretation of Denham et al. means that Farrer's
simpler data [104] taken specifically on type Ib diamonds must be
explained away. The simpler course is to take the type Ib ionization
energy as 1.7 eV.

Dean [111] obtained 4 eV for the binding energy of the donor
active in the donor-acceptor pair spectra in all types of diamond
(Sec. IX). He argued that this represents a highly localized center,

the type Ib nitrogen is a highly localized center, and therefore "it
is tempting to identify the pair spectrum donor with isolated substitutional nitrogen." Unfortunately, Dean's paper dates from 1965
when the platelet nitrogen hypothesis (Sec. III.E) was dominant,
and it seemed that the only dispersed form of nitrogen was the
type Ib form. It now appears that we should replace Dean's idea of
paramagnetic nitrogen atom of 4-eV binding energy by A aggregates
of 4-eV binding energy (see also Sec. III.D). These A aggregates
are known to occur in synthetic diamonds in concentrations that are
significant in luminescence studies. (In fact in diamonds synthesized at high enough temperature and pressure, the A and B bands
appear in the infrared absorption spectra [90].) Donor-acceptor pair
spectra involving the type Ib nitrogen atoms (binding energy 1.7 eV)
would lie at $h\nu > 3.5$ eV [Eq. (48)]. In a type Ib diamond, they
would be heavily quenched by dipole-dipole interactions (Sec. VII).

Type Ib diamonds are thus characterized by single substitutional
nitrogen atoms that undergo a large Jahn-Teller distortion, resulting
in a binding energy of 1.7 eV. The nitrogen absorbs infrared radiation (Fig. 10) and ultraviolet radiation (Figs. 11 and 12) as well as
being paramagnetic.

G.   Type IIb Diamond: Definition and Infrared Properties

*We define a diamond to be type IIb when its infrared
absorption is as Fig. 13.*

Large natural type IIb diamonds are rare. Less than 1% of specimens
with dimensions of a few millimeters are type IIb [23]. Tolansky's
survey [112] of smaller diamonds (about 0.25-mm thick) shows a much
higher fraction of smaller specimens are type II (i.e., type IIb or
IIa, Sec. III.J). Substantial variations occur in the yields of
different mines; for example, 70% of the colorless, small diamonds
from the Premier mine in South Africa are type II. Type IIb synthetic diamonds are readily produced by introducing Al, Be, or particularly B into the growth cell [113-115]. In spite of these different nominal dopants, the properties of all type IIb diamonds are
consistent with there being only one species of defect responsible

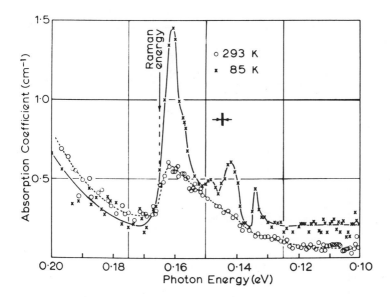

FIG. 13. Type IIb absorption at 85 and 293 K. Data by Smith and Taylor [118].

for the type IIb properties [116, 119, 120]: We discuss its identification in Section III.I.

Correlated with the strength of absorption at 159 meV, and with a peak absorption about 20 [117] to 50 times [118] more intense at room temperature, is a line at 347 meV (Fig. 14). This line has an intensity that is proportional to the concentration of neutral acceptor centers in the diamond [117, 119-121]. Collins and Williams [121] give the correlation as

$$\int A_{IIb} \ (E) \ dE = 1.27 \times 10^{-22} (N_a - N_d) \tag{15}$$

Here A_{IIb} is the absorption coefficient in mm^{-1} at photon energy E (meV). The integral is the area of the 347-meV peak with a baseline to the peak arbitrarily defined by a straight line joining the points of low absorption near 325 and 360 meV. The room temperature spectrum is used. The neutral acceptor concentration $(N_a - N_d)$ is in m^{-3}.

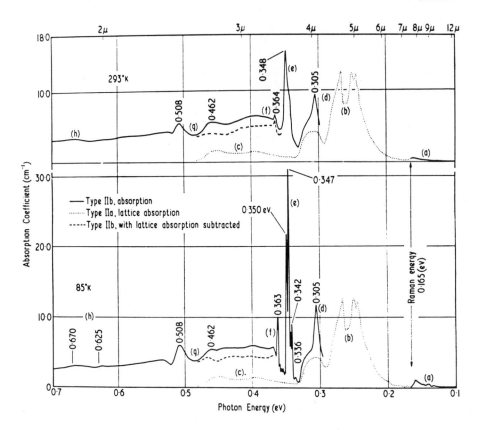

FIG. 14.   Type IIb absorption between 100 and 700 meV at 80 and 293 K.   Data by Smith and Taylor [118].

The characteristic property of type IIb diamond is thus that they are p-type semiconducting as a result of an excess of acceptor centers over all compensating defects [122].

The ionization energy of the acceptor has been established at [121]

$$E_a = 368.5 \pm 1.5 \text{ meV}$$

This low ionization energy implies that the acceptor is relative effective and mass-like [123].   Standard effective mass theory writes the hole wavefunction $\psi(r)$ in terms of valence band states $\phi_j(r)$ at $k = 0$ perturbed by the potential $-e^2/4\pi\varepsilon r$ of the hole when a distance r from the acceptor.   The hole then orbits the acceptor with

wavefunctions which are essentially Bloch states at any one point but
are localized on the acceptor by hydrogen-like wavefunction envelopes
$F_j(r)$:

$$\psi(r) = \sum_{j=1}^{6} F_j(r)\phi_j(r)$$

The summation is over the six valence band states. The Bohr radius
r_0 of the ground state is enlarged from the hydrogen Bohr radius r_B:

$$r_0 = \left(\frac{m}{m^*}\right)\left(\frac{\varepsilon}{\varepsilon_0}\right) r_B$$

as a result of the Coulomb potential being weakened by the state di-
electric constant ε. For diamond, $\varepsilon \sim 5.7\varepsilon_0$, and the hole effective
mass $m^* \sim 0.7$ m [124], so that $r_0 \sim 0.4$ nm. This is a very small
radius compared with those predicted in Si (2 nm) and Ge (4.5 nm),
and implies that the acceptor ground state $\psi_0(r)$ is going to be difficult to describe accurately since the central cell correction will be
large (especially since $\psi_0(r)$ has hydrogen 1s-like envelopes $F_j(r)$
which peak at the acceptor) and the static dielectric constant will
only just be applicable. In addition, the three valence bands are
almost degenerate (6-meV spin-orbit splitting). An attempt to calculate the electronic structure of the acceptor using effective mass
theory has been made by Bagguley et al. [138], with the valence bands
defined by hole masses and spin-orbit splitting similar to Rauch's
cyclotron results [124]. The calculation concentrated on the 2p-like
levels since these would be expected to produce the strongest absorption structure, by analogy with the hydrogen atom's having a 1s to 2p
transition many times stronger than any other transition. The 2p
levels are expected to be sufficiently delocalized that effective
mass theory may be applicable (the hydrogenic 2p radius in diamond
would be ~ 0.8 nm). Bagguley et al. found they could fit the energies
of these states very closely, but the degeneracies of some states are
probably in disagreement with Stark [139] and Zeeman [138] experiments on the acceptor center. A necessarily less-detailed, extended
Hückel calculation of the acceptor (using boron as the impurity,

Sec. III.I) is of interest in that it does not produce any mid-band-gap levels [101]. Fortunately, it turns out that the interesting properties of the acceptor can be understood using the simple ideas of effective mass theory; in fact, it is as a result of the small ground state Bohr radius alone that some features are observed in the spectra of type IIb diamonds which are only weakly observed in other semiconductors.

When discussing the type IIb absorption, it is convenient to think of three spectral regions: (a) up to 165 meV where one-phonon, infrared absorption takes place, (b) from 300 to 370 meV where internal transitions at the acceptor produce sharp line spectra, and (c) above 370 meV where the acceptor is being ionized.

Up to 165 meV, the Raman energy of pure diamond, the neutral acceptors absorb with the production of single phonons (Fig. 13). In particular, modes near 160 meV are excited reflecting the high density of optical modes [21, 22]. In contrast to the one-phonon region in type I diamonds (Secs. III.B and III.F), the absorption in a type IIb specimen is strongly temperature dependent (Fig. 13). It is also relatively strong. Using the data of Smith and Taylor [118] and Collins and Williams [121] the absorption at 159 meV in mm^{-1} is

$$293 \text{ K:} \quad A_{IIb}(159 \text{ meV}) \sim 4.7 \times 10^{-25}(N_a - N_d) \qquad (16a)$$

$$85 \text{ K:} \quad A_{IIb}(159 \text{ meV}) \sim 1.2 \times 10^{-24}(N_a - N_d) \qquad (16b)$$

At 85 K a nitrogen atom in the A form, therefore, absorbs about 60 times less than a neutral acceptor center.

The type IIb absorption is not only anomalously strong for diamond, similar one-phonon transitions are not seen in Si or Ge. The differences may be traced to the small Bohr radius of the acceptor in diamond together with its low binding energy compared with the optical mode energies ($E_a \sim 2\hbar\omega_{Opt}$). The small radius, ~ 0.4 nm, results in relatively strong coupling to optical modes of short wavelength since the coupling of a 1s-like acceptor ground state $\psi_{1s}(r)$ to optical modes of wavevector $\underline{q}$ depends on [Eq. (8) of [124]]

$$\int \psi_{1s}^{*} \; (r) \; \exp(i\underline{q}.\underline{r}) \; \psi_{1s} \; dr \tag{17}$$

The coupling makes the hole's orbital respond to the nuclear posi-
tions. But the hole's orbital frequency is only twice the vibrational
frequency, so that, in a classical picture, each orbital is continually being modified by the vibration. The coordinates of the hole
and nuclei cannot then be separated in a Born-Oppenheimer product,
but instead the nuclear motion has a large electric dipole associated
with it as a result of the continually changing hole orbital. Optical
transitions between different hole states are strongly allowed, as is
discussed later, and the mixing of the hole and nuclear motions produces the usual quantum mechanical borrowing of intensity by the
weaker system, in this case the nuclei. A detailed treatment has
been given by Hardy [126] and, as far as can be judged for this
complicated center, the enhanced one-phonon infrared absorption
arises from this breakdown of the adiabatic approximation.

There is little type IIb absorption between 165 and 300 meV, but
then sharp absorption lines are seen, especially at low temperature
(Fig. 14), as a result of internal transitions at the acceptor. Some
20 lines have been resolved by Charette [4] and by Smith and Taylor
[118] between 330 and 370 meV. Most of the lines have widths of
$\sim$1 meV at low temperature in a diamond of low acceptor concentration
$(5 \times 10^{22} \; m^{-3})$ [4, 118, 127]. As the acceptor concentration increases, interactions between centers broaden the lines until they
are almost indistinguishable from the background at $5 \times 10^{24} \; m^{-3}$
[128]. In contrast, the 0.305-eV peak is 6-meV wide even at 80 K in
a diamond of small acceptor concentration [118, 127]. Smith and
Taylor [118] suggested this is a result of lifetime broadening by
phonon coupling of the final state of the 305-meV line to another
internal state, which need not be optically active. Using a simple
theory due to Kane [129], which depends strongly on indefinables
such as the Bohr radii of the coupled states, they estimated the
coupled state to be 266 meV above the ground state. Curiously, independent evidence for a state near 266 meV turned up several years

later. Collins et al. [128] observed a weak absorption line at
268 meV in heavily doped synthetic type IIb diamond and then found
a minimum in the photoconduction spectrum which again pointed to a
266-meV state [130]. The process involved here is described in detail toward the end of this section. The photoconduction work suggested the presence of two other optically inactive states, at 240
and 289 meV [130]. (In terms of effective mass theory, states can
be present which are not observed in the absorption spectrum, just
as 1s to 2s transitions are not seen in hydrogen.)

The response of the internal transitions to uniaxial stresses
has been studied by Crowther et al. [131]. They used the effective
mass approximation as a guide to the irreducible representations of
the different hole states. Standard group theory methods could then
be used to write the response of the internal transitions to uniaxial
stresses in terms of the deformation potentials of the valence band.
A reasonable fit to the experimental shift rates resulted. This
paper contains what is probably the most accurate table of the
energies of the internal transitions, although there is some doubt
about the nature of the 2.2-meV ground state splitting [139].

We have seen that the localized ground state couples to optical
modes. The less localized excited states are less strongly coupled
according to Eq. (17). As in Sec. V.A, phonon sidebands result.
The strongest sideband at 508 meV (Fig. 14) is a 161-meV phonon
sideband of the 347 meV internal transition with an intensity about
six times smaller. This involvement of phonons of $k \neq 0$ is another
consequence of the localized ground state of the acceptor. To obtain a Bohr radius of 0.4 nm requires a spread in wavevector from
$k = 0$ to about one-sixth of k_{max} (as in Eq. 5.39 of [123]).

The third type of transition seen in the absorption spectrum of
the acceptors is the continuum beginning near 370 meV and seen with
decreasing strength through the visible spectrum [48]. This is
caused by acceptor to valence band transitions of the hole and results in type IIb diamond being gray-blue for low acceptor concentrations (5×10^{22} m^{-3}) to dark blue or opaque at high concentrations [113]. Since the valence band is $\sim$30-eV wide [132, 133], the

absorption probably continues beyond the fundamental absorption edge.
Superimposed on the continuum are the one-phonon sidebands of the
internal transitions. These peaks appear to dip below the smoothly
varying continuum (Fig. 14, see also Fig. 2 of Smith and Taylor
[118]), as though their final states are in antiresonance with the
underlying valence band states [134].

The photoconduction associated with the acceptor to valence
band hole transition has been studied in detail [135], especially by
Collins et al. [130, 136]. The photoconduction broadly follows the
absorption spectrum, but with additional features brought about by
holes of certain energies E_h having short lifetimes in the valence
band [135]. These energies, measured from the ground state of the
acceptor, are

$$E_h = E_f + \sum_{i=1}^{n} \hbar \, \omega_i \tag{18}$$

where E_f is the energy of a state in the acceptor center to which
the hole can return by rapid emission of n phonons of energy
$\hbar \, \omega_1, \hbar \, \omega_2, \ldots, \hbar \, \omega_n$. Since the valence band lies 368-meV above
the acceptor ground state [121] and the maximum phonon energy in per-
fect diamond is only 165 meV [21], the final states E_f in the capture
process are excited states of the acceptor. These excited states are
relatively diffuse; for example, the first excited state has a radius
of 0.8 nm in a hydrogenic model, and so they can be constructed from
valence band states near k = 0 (within one-sixteenth of k_{max}:
Eq. 5.39 of [123]). Consequently, phonons of k $\sim$ 0 dominate in the
one-phonon decay and photoconduction minima occur, from Eq. (18),
at phonon energies h ν = e_F + 165 meV. The first of these minima
occurs at 405, 431, and 454 meV [130] and so point to internal levels
at 240, 266, and 289 meV (Fig. 15). The next minima, labeled d and
e on Fig. 15, arise from decays of the hole, with emission of one
phonon, the decays terminating in the 304-meV level and in the
347-meV multiplet, respectively. Two-phonon emission minima (b',
c', d', e' on Fig. 15) and higher phonon processes are progressively
more diffuse as phonons of energy less than 165 meV are emitted.

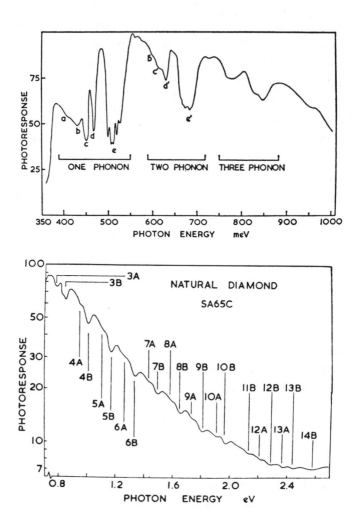

FIG. 15. Photoconductivity of type IIb diamond between 350 meV and 2.6 eV. At top, photoconduction minima are seen as a result of the reduced lifetime of the hole in the valence band corresponding to one, two, and three phonons of 165 meV, being emitted in the decay to the bound state of the acceptor. At bottom minima correspond to 3 to 14 phonons being emitted, the bound states being at 304 meV for the minima labeled A and near 348 meV for those labeled B. Data taken at liquid nitrogen temperature by Collins et al. [130].

This spread in phonon energies follows from a relaxation of the
"selection rule" that the phonon wavevector should be k ∿ 0, since
the hole can deexcite while in the valence band by emitting a phonon
of wavevector $\underline{k}$ and changing its own momentum by $-\underline{k}$. In spite of
the broadening, Collins et al. [130] have traced the sequence of
minima well into the visible spectrum (Fig. 15).

In addition to the photoconduction obtained by ionizing the
acceptor directly, it is possible to observe photoconduction from
internal optical transitions at the acceptor, for example at 347 meV,
if there is sufficient thermal energy present to ionize the acceptor
from its excited state (Fig. 16; [136]). The probability of thermal
ionization decreases rapidly with increasing energy difference between the photo-excitation state and the valence band, but apparently
all the internal absorption transitions have been observed in photoconduction with the exception of the 305-meV line. This line will be
difficult to observe because of its large separation from the valence
band (63 meV) and also because of the short lifetime of its excited
state, apparent from its 6-meV linewidth (= 10^{-13} sec).

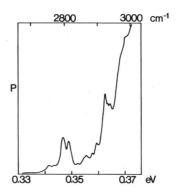

FIG. 16. Photoconduction spectrum for light incident on a
type IIb diamond below the ionization energy of 368 meV, photothermal
ionization. Spectrum measured by Collins and Lightowlers [136] at
120 K. Comparison with the absorption spectrum, Fig. 14, shows
how the photo-thermal ionization process favors internal states
nearer the ionization energy because of the limited thermal energy
available.

One particularly useful aspect of this two-state photothermal ionization process is that the photoconduction spectrum obtained is more detailed than the absorption spectrum of the same diamond [136]. Collins and Lightowlers put this down to the lifetime of a hole in an excited state of an acceptor being reduced below the time required for thermal ionization if the acceptor is in a highly perturbed part of the crystal. The photoconduction spectrum is then dominated by the unperturbed acceptors and so shows sharp lines; in contrast, the absorption spectrum does not have this built in preference for centers of long lifetime and so shows only broad bands. This qualitative explanation fits the observations [136]. Figure 17 compares the absorption spectrum with the photoconduction spectrum of two similar type IIb synthetic diamonds. The photoconduction peaks have been emphasized by subtracting a rather arbitrarily chosen background from the measured spectrum [136], but the sharpening of the structure is unambiguous.

One final infrared property of type IIb diamonds is that they absorb at long wavelengths (h ν < 30 meV) with a strength which increases with increasing temperature [137]. Clark et al. have suggested that this may arise from free carrier absorption, with the holes being excited within one valence band [137]. As such, the absorption is a property of perfect diamond rather than type IIb diamond, but the large energy gap (5.5 eV) precludes its observation in pure specimens.

H. Type IIb Diamond: Ultraviolet Properties

We have already seen that type IIb diamonds absorb visible and ultraviolet radiation, the acceptor center becoming ionized in the process. A more important ultraviolet transition is the luminescence observed after excitation by a beam of low-energy electrons (E $\sim$ 50 kV) or by photons of h $\nu \gtrsim$ 5.5 eV. Part of the luminescence, Fig. 18, arises from the radiative decay of excitons that are weakly bound to neutral acceptor centers. No detailed theory of this

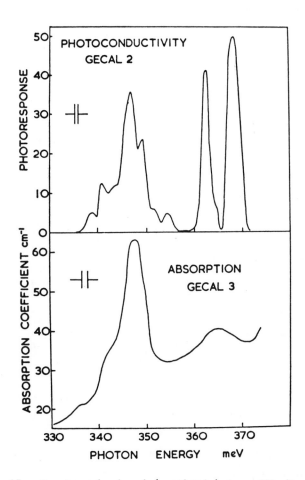

FIG. 17. In strongly doped (synthetic) type IIb diamonds the
internal transitions at the acceptor are broadened when seen in an
absorption spectrum (at bottom) but remain relatively sharp in the
photoconduction spectrum (at top). As in Fig. 16, the photoconduction spectrum is distorted by those states nearer the ionization
energy being more readily thermally ionized than more tightly bound
states. Data taken by Collins and Lightowlers [136] at 120 K. An
arbitrary baseline has been used in obtaining the photoconduction
spectrum.

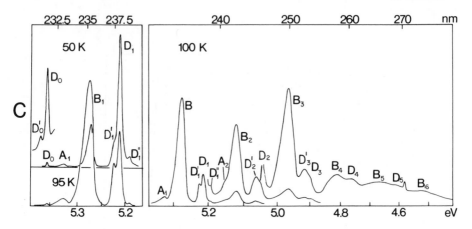

FIG. 18. Cathodoluminescence spectrum near the fundamental edge of a type IIb diamond. Features labeled A_1, B_1, C_1 are intrinsic to diamond and are caused by the decay of free excitons with the emission of one wavevector-conserving phonon. Multiphonon intrinsic components such as B_2 involve the further emission of one phonon of 165 meV ($k = 0$). See Sec. IV. Features labeled D_n are caused by the decay of excitons bound to the neutral acceptor center. The number of phonons emitted is given by the subscript. See Sec. III.H. Data by D. R. Wight, Ph.D. Thesis, University of London, 1968, taken from a natural type IIb diamond at left, at 50 and 95 K, and a synthetic type IIb diamond at right, at 100 K.

luminescence has been worked out. In addition to the complications brought about by the small spin-orbit splitting of the valence band, there is also the problem that the effective mass of the electron in the six conduction band minima is unknown. In outline, the luminescence arises as follows.

In a perfect diamond, free electrons and holes may combine to form a Wannier exciton. Taking a naive view, the binding energy of this exciton will be

$$E_x \sim R_H \left(\frac{m_x}{m}\right)\left(\frac{\varepsilon_0}{\varepsilon}\right)^2 \qquad (19)$$

where

$m_x = m_e m_h / (m_e + m_h)$ = exciton reduced mass
R_H = Rydberg constant for hydrogen

Decay of the exciton corresponds to the electron making a transition between states that are derived from the six conduction band (CB) minima at k = < 0.76 k$_{max}$, 0, 0 > and from the valence band (VB) maxima at k = 0. The change in electron wavevector must be balanced by involving phonons of the same magnitude of wavevector. For example, the features B$_1'$ and B$_1$ (Fig. 18) with thresholds at 5.275 and 5.268 eV involve emission of transverse optic modes of 141 meV. Free excitons of zero translational energy, therefore, have energies 5.416 and 5.409 eV. Dean et al. [116] associate the 5.416 eV exciton with holes derived from the lower (split-off) VB and the 5.409-eV exciton with the upper VB. The difference in energies of these states is then partly due to the 6-meV VB spin-orbit coupling [124] and also to different hole masses in the two sets of bands, since these change E$_x$ through m$_x$ [Eq. (19)]. To estimate the mass effects, we note that the upper VB is 5.49 ± 0.005 eV below the CB minima [140], so that the upper VB exciton has a binding energy E$_x$ $\sim$ 5.49 − 5.409 eV = 81 meV. Taking the hole mass as the mean of the two upper VB values, 0.7 and 2.1 m [124], gives from Eq. (19) m$_e$ = 0.225 m. Unfortunately, there is no direct measurement of the CB minima electron masses for comparison with m$_e$. However, using m$_e$ with the split-off VB mass m$_h$ = 1.06 m [124], we find E$_x$ = 77.5 meV. Thus we could naively expect luminescence from this lower VB exciton at an energy 9.5 meV above the upper VB exciton's luminescence (9.5 meV includes 6 meV for spin-orbit coupling). The experimental value is 7 ± 1 meV. The calculation is not to be taken seriously, but it does illustrate that hole mass effects are not negligible, contrary to the assumptions implicit in Ref. 116.

The mass effects become more significant when the exciton is bound to a neutral acceptor, since the binding energy is mass dependent: Qualitatively it always becomes more difficult to localize a particle as its mass decreases [141a]. Hopfield has calculated the binding energy for the simplest case of an effective mass exciton bound to a neutral effective-mass-like donor (or acceptor) for nondegenerate valence and conduction bands [141b]. Using the effective

masses derived previously, the exciton binding energies are predicted
to be about 18 meV. The observed values, based on the energies of
the zero-phonon lines D_0' and D_0 (Fig. 18) are 53 and 48 meV for
the upper and lower VB excitons [116]. The simple theory gives the
correct order of magnitude, and so confirms that neutral acceptors
are involved, but it is not sufficiently accurate to give a con-
fident prediction of the change in binding energies resulting from
the mass changes. However, we can expect the lower VB exciton to
have a binding energy to the neutral acceptor, which is a few millielectronvolts less than the binding energy of the upper VB exciton,
as observed. This would seem to be a more satisfactory explanation
of the change in binding energy than postulating [116] the spin-orbit
coupling is doubled near the acceptor, which is probably the light
element boron (Sec. III.I).

The vibronic properties of the luminescence can be rationalized
qualitatively in terms of the structure of the center. The complex
of an exciton bound to a neutral acceptor contains one electron and
two indistinguishable holes. The hole provided by the acceptor is
relatively tightly bound (Sec. III.G) and so the radiative decay
shows some zero phonon components [116] (D_0' and D_0 in Fig. 18).
However, the bulk of the decay occurs with the emission of transverse
optical phonons of wavevector $< 0.76\ k_{max}$, 0, 0 >, approximately;
these phonons conserve momentum in a free exciton decay, and we can
say that their emission reflects the loosely bound exciton properties
of the complex. These phonon sidebands are labeled D_1' and D_1 on
Fig. 18. They are some 62 ± 8 times stronger than the zero phonon
lines ($D_0' + D_0$) [116]. Multiphonon sidebands (D_2', D_2; D_3' D_3)
involving Raman phonons (165 meV) as well as the transverse optical
ones are also present. No absorption corresponding to these luminescence transitions has been detected, although these are known in
other materials, e.g., GaP:Zn [142].

So far we have considered neutral acceptors. Excitons may be
bound at ionized acceptors if the masses of the electrons and holes
are suitable: $m_e/m_h > 1$ in Hopfield's simple theory [139]. Using
$m_e = 0.225\ m$, derived before, and $m_h \gtrsim m$, we see that this condition

will not be met. But different, and equally valid, numerology by
Dean et al. [116] suggests excitons could be bound to ionized acceptors long enough to decay radiatively, as long as the temperature is
sufficiently low (T < 25 K). The necessary measurements have not yet
been made; at the moment it is debatable whether a high enough concentration of ionized centers could be produced. Figures of 99%
compensation of the acceptor centers, which would give high concentrations of ionized centers, are quoted by Dean et al. [116], but
these are based on an incorrect interpretation of chemical analyses
(Sec. III.I). It now seems that the compensation is only of the
order of a few percent of the acceptors in most natural [121] and
in good synthetic [117] semiconducting diamond.

The high concentration of neutral acceptors suggests that luminescence may be observed from free electrons recombining with holes
in the acceptor ground state. The luminescence would have a threshold at $h\nu = E_g - E_a \sim 5.122$ eV at 100 K for its zero phonon component, which would be allowed as another consequence of the localized
acceptor ground state (Sec. III.G). The shape of the zero-phonon
line would be expected to be similar to that of the free-exciton
peaks, Eq. (20d), such as B_1 on Fig. 18, since the free electron
would have a range of thermal energies in the conduction band minima.
Zero-phonon lines of widths $\sim kT$ would be expected. Experimentally,
a peak is observed with a threshold at 5.135 eV at 100 K in some
type IIb diamonds, and this has been suggested to arise from this
free-electron capture at neutral acceptors [116]. However, the peak
has also been detected in type IIa diamond [183], and in general,
there does not seem to be any relation between the strength of the
5.135-eV line and the D lines of Fig. 18, which are also interpreted as involving neutral acceptors. The origin of the 5.135-eV
line must still be regarded as undetermined. The line can be surprisingly wide, e.g., 25 meV at 95 K in one type IIb diamond [116]
and 50-meV wide in another good type IIb specimen. There is a
165-meV phonon sideband of about half the integrated strength of the
main line.

I. Type IIb Diamond: The Nature of the Acceptor

By analogy with silicon and germanium, we expect that when a
Group III element substitutes for a carbon atom in a diamond the
impurity will form an approximately effective mass-like acceptor
[123]. Consequently, the acceptor center is most likely either B,
Al, Ga, or In. Chemical analysis [8b] shows Ga and In to be present
in much too small a concentration to account for the number of
acceptors measured from Hall effect studies. Aluminium is present
in most diamonds (Table 1), but as we saw in Sec. II, it is also an
important constituent of inclusions. Analysis of five carefully
chosen natural type IIb diamonds by Collins and Williams showed that
in each specimen the Al concentration was less than the total number
of acceptors [121]. The Al concentration was measured using the
reaction $^{27}Al(n, \gamma)^{28}Al$, which probes all the diamond. Bulk measurements of the acceptor concentration were made from the Hall
effect. Therefore we can safely rule out Al as the acceptor center
in diamond. This result has been confirmed by Chrenko using synthetic diamonds [117].

We are left with boron as the only remaining Group III candi-
date. Chemical analysis for B concentrations in the parts per mil-
lion range requires nuclear reactions involving charged particles,
such as either $^{10}B(n, \alpha)^{7}Li$ [114] or $^{11}B(p, \alpha)^{8}Be \to 2$ [143], and
so the experiments only monitor the region 1 to 2 μm into the crystal
surface. Chrenko [117] has compared the B concentration, measured
by the ^{10}B-neutron reaction, with the number of uncompensated acceptors $N_a - N_d$, measured using the 347-meV absorption bandstrength
[Eq. (15)]. His data are consistent with B being the acceptor, the
large scatter in his results probably arises from comparing defect
concentrations measured in different regions of the diamonds. However, a definite identification of the acceptor will require measurement of the acceptor concentration in the same surface layer as used
for the B analysis. The photocapacitance technique would be suitable [144].

The current situation is thus that substitutional B appears most likely to be the acceptor. This is not proved, but is consistent with many pieces of information about type IIb diamonds. For example the deep blue color of synthetic diamonds is only produced on deliberate doping with B [113], and the one-phonon, infrared spectrum (Fig. 13) is of the shape we would associate with a substitutional atom of mass very similar to carbon so that the spectrum can reflect the high density of optical modes of diamond [21, 22].

Unfortunately the literature is confused by frequent references to the acceptor center being Al, especially in papers dating from 1965 to 1970 [8b, 111, 116, 128, 131, 136, 145-149]. This misidentification [150] arose largely from the Al concentration in three natural type IIb diamonds being almost equal to the acceptor concentrations in the same specimens [116]. Lightowlers and Collins found that much experimental data could be quantitatively understood on the assumption of a large range of compensation of the supposed Al acceptors [111, 145, 148]. Early work on the synthesis of type IIb diamond did little to clear up the error, since important acceptor concentrations ($\sim 10^{22}$ m^{-3}) can result from low levels of B contamination of the growth cell. Ironically, in the paper announcing the synthesis of p-type diamond, Wentorf and Bovenkerk reported that it was Al doping that produces the standard acceptor level with a binding energy of 0.32 eV, according to their measurements [113]. Boron doping gave an apparent binding energy of only 0.17 to 0.18 eV, presumably because the crystals were near degenerate. Only later was it suggested that the Al played the role of a getter for nitrogen, permitting accidentally present B to give semiconducting diamonds.

The misidentification of the chemical nature of the acceptor center has almost no consequences in the optical studies apart from the need to read "the acceptor is Al" as "the acceptor is probably B." In some papers [e.g., 116, 151], optical features were tentatively assigned to B acceptors that were supposed to be compensated by the dominant Al acceptors; for example, the 5.26-eV cathodoluminescence peak is suggested to be due to exciton recombination at an

ionized B acceptor (line S_0 in Ref. 116). These ideas must now be
ignored.

It is disturbing to realize that inaccurate models have been
presented for both major impurities in diamond, the acceptor center
in type IIb diamond and the nitrogen in type Ia diamond, and that
these models have been accepted in numerous publications without
their ever having had concrete foundations.

J. Type IIa Diamond and the Limitations
of Type Classification

So far we have made unambiguous definitions of type IaA, IaB,
Ib, and IIb diamonds using the shape of the absorption spectrum of
infrared radiation in the one-phonon region. In the limit of small
nitrogen (or acceptor) concentrations, the absorption becomes neg-
ligibly small, and the diamonds are classified as type IIa.

*A type IIa diamond has no detectable infrared absorp-
tion in the range $h\nu < 1332$ cm^{-1}.*

The key word is "detectable." For example, consider a type IaA diamond 2-mm thick. To detect the presence of nitrogen, we need at
least a few percent absorption of infrared radiation of 1282 cm^{-1}.
Consequently, if the diamond has a nitrogen concentration of less
than 10^{24} m^{-3}, we would classify it as type IIa [see Eq. (3)]. But
now if a few percent of the nitrogen is converted into a form that
gives electric dipole allowed transitions, for example by reactions
with radiation damage products, then the diamond will begin to appear
much as an irradiated type Ia diamond. A specific example is the
frequent occurrence of the 3.149-eV (ND1 or R10) absorption band
in type IIa diamond after irradiation with damage creating particles
(Sec. V.B). Similarly, since in a diamond of low nitrogen content,
optical centers usually have a very high luminescence efficiency,
it is often easy to observe the 2.463-eV(H3) and the 2.985-eV(N3)
luminescence bands from an untreated type IIa diamond, just as from
a type Ia diamond [48].

The reason for these weak type Ia properties is that the nitrogen content in type IIa diamonds is, in the few specimens analyzed,

about 6 ppm or 10^{24} m^{-3} [3, 6, 11]. This is at least two orders of magnitude higher than the nitrogen level in type IIb diamonds: Type IIa diamonds are not the purest type. The type IIa classification is thus doomed to obsolescence as soon as a higher sensitivity of detecting A and B nitrogen aggregates becomes available. A typical type IIa diamond has continuous absorption in the ultraviolet (Fig. 7). The shape of this continuum is the same for all type IIa samples [48] and is also similar to the continuum produced by annealing radiation damage [152].

K. Intermediate Type Diamonds

If the type IIa classification is heading to obsolescence, the intermediate classification has already got there, and it is only mentioned here in view of its frequent occurrence in the literature. Clark et al. [48] introduced the term to describe those diamonds with a low ultraviolet absorption, which was thought to imply type II properties, in diamonds showing those vibronic bands such as the 2.985-ev(N3) band, which are seen in type I diamonds. Later the term came to imply diamonds of low ultraviolet absorption but nonzero, one-phonon absorption [27], and it became clear that this class of diamonds contains nitrogen predominantly in the B form. Since we can define type IaB diamonds with precision (Sec. III.B), we do not use the loose term "intermediate diamonds" in this chapter.

The main interest in this type of diamond stems from the possibility of making optical measurements near the fundamental absorption edge, since there is little extrinsic absorption there, and watching the interactions of the free charges with the 5.26-eV(N9) optical center, which always accompanies the B nitrogen [49, 148].

IV.   THE OPTICAL PROPERTIES OF PURE DIAMOND

From an optical point of view, pure diamond does not exist in nature. To deduce the properties it would have, we must seek the features common to all the different types of diamond, concentrating especially on the relatively pure types IIa and IIb. In this section, we outline these properties, working from low to high photon energies.

A pure diamond would have a negligible concentration of electrons excited from the valence band to the conduction band at room
temperature: Negligible here means less than one electron per cubic
meter. Consequently, there is zero absorption due to electronic
excitation until $h\nu \sim 5.5$ eV. The absorption in the infrared is the
result of two or more phonons being created; single phonon creation
is forbidden in perfect diamond, as discussed in Sec. III.A.
The absorption becomes appreciable above 1400 cm^{-1} (Fig. 1). Up
to about 2350 cm^{-1}, it is easy to distinguish the two-phonon band
from any superimposed structure such as the $\sim$1370 cm^{-1} line
(Sec. III.C) or the very sharp lines at 1405 cm^{-1} [4, 11, 40, 153],
1520 and 1540 cm^{-1} [25, 153] seen in some type I diamonds. The
two-phonon band increases in strength to 2000 cm^{-1} where there is a
small dip. At room temperature the absorption at the bottom of this
dip is 1.23 mm^{-1} for all diamonds: This provides a convenient normalization point when measuring the one-phonon region.

As yet there is no complete theoretical fit to the shape of the
two-phonon spectrum. Normal mode frequencies and eigenvectors can
be calculated, within any given lattice dynamics model, by a fit to
the lattice dispersion curves measured by inelastic neutron scattering [154, 155]. The two phonon bandshape can then be calculated if
the local dipoles introduced by the nuclear displacements can be
estimated [20]. Initial calculations [156] took into account only
the nearest neighbor atom interactions and so could not fit the
observed spectrum [157]. Most of the structure is reproduced by the
seven parameter model of Wehner et al. [21]. Solin and Ramdas [158]
have made a serious attempt to rationalize the infrared absorption
data, the second-order Raman effect, and the dispersion curves of
diamond (see Table 2). Their approach, which is based on the selection rules for the optical properties, gives the phonon critical
points more accurately than neutron scattering, but it cannot predict the spectral shapes. The present situation, then, is that a
complete understanding of the two-phonon spectrum has not been obtained.

TABLE 2

Phonon Energies at High Symmetry Points in the Brillouin Zone
(from a comparison by Solin and Ramdas of optical
and neutron scattering work [158])

Symmetry point	Phonon	Energy cm^{-1} (meV)[a]	
Γ	TO,LO	1332 ± 0.5	(165)
X	TO	1069	(132)
	LA,LO	1185	(147)
	TA	807	(100)
L	TO	1206	(149)
	LO	1252	(155)
	TA	563	(70)
	LA	1006	(125)
W	TO	999	(124)
	LA,LO	1179	(146)
	TA	908	(112.5)

[a]Uncertainty is ±5 cm^{-1} except for Γ.

Before the dispersion curves were measured, there were three
attempts to derive them from optical data. Hardy and Smith [159]
analyzed the two-phonon spectrum, while Johnson and Loudon [160]
and Bilz et al. [161] concentrated on the second-order Raman effect.
These analyses are not unambiguous [160], and in fact, the results
were qualitatively different from the dispersion curves as measured
later [154].

With increasing temperature, lattice vibrations are increasingly
present in the diamond and additional infrared absorption paths become available. For example, the two-phonon spectrum can now con-
sist of one phonon being absorbed and one emitted (Eqs. 5.22 and
5.23 of [20]). Surprisingly, there are no experimental data available about this.

If a transition can occur between two states with the absorption
of a photon, the reverse transition will also be allowed. Luminescence in the two-phonon region is readily observed by heating a
diamond to about 700 K and measuring its thermal radiation [37].

Kirchhoff's law of emissivity requires this luminescence to have an intensity

$$I(\nu) = B(\nu)[1 - \exp(-\mu(\nu)t)]$$

where $B(\nu)$ is the emission of a black body whose temperature is the same as the diamond's, and the bracketed term is the fraction of infrared radiation absorbed by the diamond. This relationship is observed experimentally [37].

The rule that no absorption can occur with the creation of one phonon is lifted if the lattice inversion symmetry is broken [162]. In practice, electric fields of the order of 10^7 V m^{-1} are required to induce measurable absorption which then consists of a line at the Raman frequency, since the phonons must have $\underline{k} \sim 0$ to conserve wavevector [163]. Similarly, the two-phonon absorption band may be changed by an electric field, although the fraction change in absorption is only of the order of 10^{-4} with fields of 10^7 V m^{-1} [164].

Perfect diamond is transparent to radiation of photon energies above about 3996 cm^{-1}, the cut-off of the three-phonon absorption, until electron band effects become important near 5.5 eV. Similarly, perfect diamond is nonluminescent in this range, and the only known optical property is Raman scattering [36, 165-176]. Diamond has the ^7O_h space group and so the three optical modes are degenerate at $\underline{k} = 0$. The first-order Raman spectrum therefore consists of one line which, at room temperature, is displaced 1332 cm^{-1} from the exciting frequency. The linewidth is about 1.9 cm^{-1} at room temperature, increasing with temperature to about 3.5 cm^{-1} at 900 K [158, 171, 174, 175]. The width is solely due to lifetime broadening of the Raman phonon; for example, at room temperature, the lifetime is (2.9 ± 0.3) x 10^{-12} sec, corresponding to a linewidth of 1.8 ± 0.2 cm^{-1} [177]. These lifetime effects come from the anharmonic vibrational potential which permits the Raman phonon to decay into two acoustic modes [174, 177]. Anharmonicity also changes the Raman energy from 1333 cm^{-1} in the limit of low temperature [174] to about 1316 cm^{-1} at 1150 K [172]. The shift appears to be identical with that of the

1332 cm^{-1} resonance seen in the one-photon spectrum of the B nitrogen aggregates (Sec. III.B). We may write the Raman wavenumber $\bar{\nu}$,
at constant pressure, as a function of the diamond's temperature
and volume, giving

$$d\bar{\nu} = \left(\frac{\partial\bar{\nu}}{\partial T}\right)_V dT + \left(\frac{\partial\bar{\nu}}{\partial V}\right)_T dV$$

The volume derivative is known, essentially, from room temperature
hydrostatic stress experiments, which show the Raman frequency
changes linearly with applied pressure P (pascals):

$$\bar{\nu} = 1332 + 0.29 \times 10^{-8} \text{ P cm}^{-1}$$

This implies a Raman Grüneisen parameter $\gamma = 1.1 \pm 0.2$ [173]. The
volume dependence is then $(\partial\bar{\nu}/\partial V)_T \Delta V = \bar{\nu}(T = 0)[\exp (\gamma \Delta V/V) - 1]$,
where $\Delta V/V$ is the lattice expansion. This static lattice expansion
term is found to describe the shift of the Raman line up to about
700 K [174]. Above this, the dynamic term $(\partial\bar{\nu}/\partial T)_V$ produces a further shift. Qualitatively, this is as expected, for the static
lattice expansion term will be adequate at temperatures low enough
that only the long wavelength acoustic modes, with their small relative displacements of neighboring atoms, are thermally populated.

The intensity of a first-order Raman spectrum depends on the
electronic polarizability per unit cell with respect to atomic displacements, $d\alpha/dx$. Direct measurement of the Raman efficiency gives
$d\alpha/dx = 4.6 \times 10^{-20} \text{ m}^2$ [175] in excellent agreement with the value
obtained from the electric field-induced, one-phonon absorption [163].
This polarizability corresponds to $(2.7 \pm 0.9) \times 10^{-8}$ photons Raman-
scattered into a unit solid angle for each millimeter path length
traversed in the diamond by the incident light beam [175]. This
figure for the Raman efficiency is independent of power up to incident energies of about 10^{13} W m^{-2}. At this value, stimulated Raman-
scattering sets in with a very rapid rise ($\sim10^6$ x) in Raman efficiency as the incident power increases to about $1.5 \times 10^{13} \text{ W m}^{-2}$
[175]. Higher intensities cause crystal damage.

The second-order Raman effect is the creation, in the Raman-scattering process, of two phonons of essentially equal but opposite wavevector. The spectrum consists of a broad-structured peak stretching from 2100 to 2690 cm^{-1} below the excitation frequency. The peak intensity, at 2460 cm^{-1}, is about 250 times weaker than the first-order peak intensity [158]. Second-order Raman spectra arise from the second derivative of the polarizability of the lattice with respect to atomic displacements in the excited modes (Eq. 49.5 of [178]). Consequently, an explicit calculation based on the normal modes of vibration must employ simplifying assumptions, and only an outline fit has been obtained. The latest calculation by Cowley is given in Fig. 10 of [158]. The main feature not reproduced by the calculation is a sharp spike at 2666.9 ± 0.5 cm^{-1} at 300 K [158, 165]. Polarization data for this line show that it is not produced by two single first-order scattering processes [158]. The two-phonon density of states function does not show a sharp peak at its high-energy end [21, 156] so that the line is probably not a simple overtone of the 1332-cm^{-1} line. Cohen and Ruvald [179] have suggested the line is caused by the creation of a two phonon-bound state. The phonons are derived from $\underline{k}$ = 0 optical modes and so have negative effective masses. Stability of the bound system then requires the anharmonic phonon-phonon force to be repulsive. Trial calculations are consistent with the experimental data [179].

Confirmation of the interpretation of features in the second-order Raman spectrum would come from studying its response to uniaxial stresses. To date these experiments have not been performed. (Hydrostatic stress measurements are in progress [180], but these give only the mode Grüneisen parameters.)

At low temperatures, diamonds cannot absorb ultraviolet photons of $h\nu \lesssim 5.5$ eV, but above this energy valence band to conduction band transitions are seen. The form of the absorption shows the band gap to be indirect [181]. At low temperature, the absorption then takes the form of a series of square-root terms, each being

$$A_i(\nu) = M_i(h\nu - E_i)^{1/2}, \quad h\nu \geq E_i \tag{20a}$$

with the absorption beginning at a threshold

$$E_i = E_g - E_x + \hbar_i \qquad (20b)$$

where E_i is determined by the electronic energy gap E_g, the exciton binding energy E_x and the particular phonon $\hbar\omega_i$ involved in the transition. These phonons must conserve crystal wavevector by balancing the change in electron wavevector when it makes the indirect transition. Clark et al. [140] observed two strong absorption terms, corresponding to phonons of $\hbar\omega_i$ = 83 ± 3 and 143 ± 2 meV. The dispersion curves of Warren et al. [154] show these to be, respectively, transverse acoustic and transverse optic phonons of wavevector < $0.7k_{max}$, 0, 0 >. The strengths of these bands were in the ratio $M_{TO}/M_{TA} \sim 8$. A further phonon of $\hbar\omega_i \sim 132$ meV was identified on poor evidence [140, 182, 183]. It was dropped in later papers [116].

The luminescence at low temperature is related to the absorption, Eq. (20), but since phonons must again be created to conserve wavevector, it is Stokes-shifted to thresholds

$$E_i = E_g - E_x - \hbar\omega_i \qquad (20c)$$

The continuous absorption, Eq. (20a), corresponds to excitons being created with different center-of-mass kinetic energies, and so is modified by a Boltzmann factor in the luminiscence spectrum:

$$I_i(\nu) = m_i(h\nu - E_i)^{1/2} \exp\left[\frac{-(h\nu - E_i)}{kT}\right] \qquad (20d)$$

Luminescence of this form is observed experimentally (A_1', A_1, B_1', B_1, C_1', C_1 of Fig. 18). Each phonon component appears twice because of the spin-orbit coupling in the exciton (Sec. III.H). Sharp thresholds at $h\nu = E_i$ are not observed as a result of inhomogeneous broadening by crystal defects [184]. The A and B peaks involve transverse acoustic and transverse optic phonons of 87 ± 2 and 141 ± 1 meV [116] in reasonable agreement with the absorption data. The phonon dispersion curves [154] give wavevectors < $0.76k_{max}$, 0, 0 > for these modes. The luminescence intensity of the B and A features is large, $m_{TO}/m_{TA} \sim 20$, as expected from the absorption spectrum where

$M_{TO}/M_{TZ}A \sim 8$. These ratios will be equal if the principle of detailed balance is applicable, that is, if we can ignore the Stokes shift of the two sets of transitions. The discrepancy probably lies in the difficulty of finding M_{TO}/M_{TA}: It involves separating many overlapping components whose thresholds are not very well defined because of the inhomogeneous broadening. A test of the applicability of detailed balance could be made using, for example, the ith Stokes component of luminescence with the ith anti-Stokes component of absorption. The available data are not sufficiently detailed to analyze this.

Multiphonon components can be observed in the luminescence spectrum. In these, one or two Raman phonons (165 meV) are created simultaneously with the wavevector conserving phonon: B_2 and B_3 on Fig. 18 are examples. Corresponding features should be present in the absorption spectrum but the data of Clark et al. [140] are unreliable, partly because of the high absorption coefficients (>100 mm^{-1}) at the relevant phonon energies and partly because of the numerous overlapping components.

All these transitions involve excitons. Free carriers can also be created according to

$$A_i(\nu) = N_i(h\nu - E_i)^{1.5}, \quad h\nu \geq E_i \tag{21a}$$

The threshold is now

$$E_i = E_g + h\omega_i \tag{21b}$$

The strongest component is again that involving transverse optic phonons with a threshold at 5.615 eV at room temperature (Fig. 7 of [140]). At sufficiently high temperature, in practice T > 400 K, excitons are thermally dissociated, and free carrier recombination begins to affect the luminescence by giving components of the form of Eq. (21) multiplied again by a Boltzmann factor (Fig. 2 of [116]). Luminescence from the condensed electron-hole phase (electron-hole-droplets) has not yet been observed in diamond. It would be expected near 5.22 eV at low temperature ($\lesssim 20$ K).

We would expect to find photoconduction when light of band gap energies is incident on a perfect diamond, particularly from the free carrier creation. This has been observed experimentally [36, 49, 50, 52]. Denham et al. [49] also report photoconduction at 90 K associated with *exciton* creation, both transverse optic and transverse acoustic components being seen. They suggest this is caused by thermal dissociation of the excitons, but there would seem to be more to it than this, since there is a rise by only about a factor of eight in this photoconductivity on heating from 90 K to 200 K (see component X on their Fig. 1) while the Boltzmann factor $\exp(-E_x/kT)$ increases by a factor of 150. One other possible dissociation mechanism, not considered in [49], is that the free carriers could ionize the excitons by being inelastically scattered from them as the carriers are accelerated through the crystal by the applied electric field, especially near the electrical contacts where probably most of the potential is dropped. In a sufficiently pure diamond at low temperatures, the carriers should reach these energies since the upper limit to their kinetic energy will be determined by optical phonon creation [185]. The limit is thus at a kinetic energy $\hbar\omega \sim 2E_x$, ample to dissociate the exciton. Further experiments on the photoconduction are required to clear this up.

The absorption coefficient for photons of higher energy increases very rapidly, being 500 mm^{-1} at 5.96 eV [140, 188], and we reach the limit of absorption spectroscopy. Reflection spectra have been taken by many workers [140, 186-190]. The upper limit to date is 31 eV [189]. The real and imaginary parts of the dielectric function have been extracted by Kramers-Kronig analysis [186-189]. These should be calculable from a model of the electron band structure of diamond: This is the main significance of the reflection measurements. The reflection rises steadily to about 7 eV, where there is a peak arising from direct transitions between the valence band and the lowest conduction band [140, 186-190]. It is not clear which points in electron wavevector space are producing this transition [191]. The highest reflection, about 60%, occurs at 12 eV and is probably not associated with any particular critical point in the

electron band structure. At higher energies, the reflection falls
again, the main point of interest being the evidence for free-electron
plasma resonance near 30 eV [189], in agreement with the 31-eV plasmon
reported by Whetton [192] from electron energy-loss measurements. In
practice, it appears then that reflection measurements are not very
helpful in fixing the electron band structure of diamond.

Increasing the photon energy still further takes us into the
x-ray photo-electric regime [133] and out of the range of this survey.

V. SIMPLE VIBRONIC SPECTRA

A. *General Theory*

So far in this chapter, we have described the properties of different
types of diamond. In the remainder, we concentrate more on the different sorts of optical process. We begin with the vibronic properties
of some of the localized optical centers. By "localized," we mean
that the optically active electrons are essentially confined within
one or two atomic spacings of a small crystalline imperfection (e.g.,
a vacancy). A trivial model of this type of center is to treat it as
having one optical electron which is completely confined in a fea-
tureless cubic region of side ℓ. Solution of the Schrodinger equation shows that the electron can have energies

$$E = \frac{h^2}{8m\ell^2} n^2, \quad n = 1,2,3,\ldots \tag{22}$$

An electron confined to be within two atomic spacings of the defect
will then give electronic transitions between its two lowest states
at E = 2.94 eV, ℓ = 0.62 nm. Consequently, we are concerned here
with transitions in the mid-band-gap region [193].

Most of the properties of these transitions arise from their
electron-lattice interactions. To get some feeling for these interactions, we pursue our simple model further. When the electron is
excited to the n = 2 state, its energy, $3h^2/8m\ell^2$, can be altered by
a change in the size of the box. Changing the size to $\ell' = \ell + Q$
gives an electronic energy

$$E(\ell') \simeq E(\ell) \left[1 - \frac{2Q}{\ell}\right] = E(\ell) - cQ \quad \text{say} \tag{23}$$

The linear electron-lattice coupling parameter c is of nonnegligible size; for example, a moderate strain of $Q/R = 10^{-3}$ will change the electronic energy by 6 meV, and this can be readily measured (Sec. VII). A more immediate effect of c is that it provides a driving force for the box to expand when the electron is in the first excited state n = 2. Physically, this expansion would occur through a movement of the atoms neighboring the optical center. The characteristic time for atomic movements, $\sim 10^{-13}$ sec, is far shorter than the time required for de-excitation of the electron by spontaneous photon emission, $\sim 10^{-8}$ sec (Sec. VII). This again makes it favorable for the lattice relaxation to occur. The extent of the relaxation is limited by the harmonic restoring force of the lattice. A distortion Q increases the potential energy by an amount kQ^2 so that the total energy of the center follows

$$\xi = E(\ell) - cQ + kQ^2$$

$$= E(\ell) + k\left[Q - \frac{c}{2k}\right]^2 - \frac{c^2}{4k}$$

The energy is still parabolic in Q so the imaginary box still vibrates harmonically but now about an equilibrium point

$$Q_0 = \frac{c}{2k}$$

The coupling has lowered the energy by an amount $c^2/4k$. For example, a 5% relaxation, $Q_0 = \ell/20$, lowers the energy by $c\ell/40 \sim E(\ell)/20$. This is comparable with phonon energies and results in phonons being involved in electric dipole transitions.

In reality, an optical center can vibrate in some 10^{29} modes per cubic meter of crystal, while we have introduced only one vibrational coordinate Q. As long as we stick to linear electron-lattice coupling and to harmonic modes of vibration that are totally symmetric in the point group of the optical center, the generalization to many modes is straightforward, for each mode vibrates independently of all others. The theory of the resulting optical spectra is well established [194-197], although it has been subjected to few rigorous tests in any crystals [198, 199].

We define a function $g(\omega)$ such that $g(\omega)\hbar \, d\omega/\omega$ is the total energy reduction as a result of linear coupling to all totally symmetric modes of frequencies ω to $\omega + d\omega$. This energy reduction was $c^2/4k$ in the simple model. The probability of the transitions that create in unit time one phonon of frequency ω is (Eq. 13.45 of [195])

$$I_1(\omega) = \frac{C \, e^{-S} [n(\omega) + 1] \, g(\omega)}{\omega^2} \tag{24}$$

where C is a constant involving the electric dipole matrix element between the ground and excited electronic state (in fact, C is the total band transition probability), and $n(\omega)$ is the phonon population factor.

The corresponding probability for the transitions that destroy one phonon takes the usual form of

$$I_1(-\omega) = \frac{I_1(\omega) \, n(\omega)}{[n(\omega) + 1]} \tag{25}$$

The probability of creating two phonons of total energy E is given by all ways of combining two one-phonon processes, since the phonons are independent in our present treatment. Thus

$$I_2 \frac{E}{h} = N_2 \int I_1 \left(\frac{E}{h} - \omega\right) I_1(\omega) \, d\omega \tag{26}$$

Similarly the probability of an n-phonon process is

$$I_n \frac{E}{\hbar} = N_n \int I_{n-1} \left(\frac{E}{h} - \omega\right) I_1(\omega) \, d\omega \tag{27}$$

The normalizing constants N_n are chosen so that the total n-phonon probability is [197]

$$\int I_n(\omega) \, d\omega = \frac{C \, e^{-S} \, S^n}{n!}, \quad n = 0,1,2,\ldots \tag{28}$$

The Huang-Rhys factor S may be readily expressed in terms of the linear coupling function $g(\omega)$ through the one-phonon transition probability:

$$\int [I_1(\omega) + I_1(-\omega)]\ d\omega = \int \frac{I_1(\omega)[2n(\omega) + 1]}{n(\omega) + 1}\ d\omega$$

from Eq. (25) and so from Eqs. (24) and (28)

$$S = \int \frac{d\omega\ g(\omega)[2n(\omega) + 1]}{\omega^2} \qquad (29)$$

At 0 K, $n(\omega) = 0$, and S is seen to be the sum over all modes of the
energy reduction resulting from linear coupling to one mode divided
by the quantum $\hbar\omega$ of that mode. Experimentally, S can be found from
the transition probability of the zero-phonon line relative to the
total band intensity [Eq. (28)]. Note that Eqs. (24) through (28)
relate to transition probability. To get an absorption coefficient,
the transition probability must be multiplied by the photon frequency
ν, and to get an emission intensity (number of photons emitted per
unit energy range) it must be multiplied by ν^3.

We see that the shape of a band at any temperature is contained
in the unknown function $g(\omega)$ and the Huang-Rhys factor measured at
one arbitrary temperature. However, the theory is limited so far to
linear electron-lattice coupling, and there is no reason why phonons
should not interact through terms $Q_i Q_j$ in the displacement Q_i of
each mode. (The analogous term in the simple box theory is
$3E(\ell)Q^2/\ell^2$, the next term in the Taylor series expansion in Eq. (23).)
As long as this quadratic coupling is small Eqs. (24) through (28)
remain approximately valid [194]. One simple test of the validity
is to see if the absorption and emission spectra of the optical
center are mirror images of each other [198]. This is the case.
For example, at the 2.985-eV(N3) center, the energy of the acoustic
modes measured from the absorption spectrum is indistinguishable
from their energy measured in luminescence. The same is true for
the optical modes. Comparisons of this type cannot be made at the
other centers discussed here. For these we note that their zero-
phonon lines broaden at similar rates to that of the 2.985-eV line
as the temperature is increased.

The most easily observed effect of quadratic coupling is this
broadening of the zero-phonon line as the temperature is increased.

The broadening takes the form of a Lorentzian function. At suf-
fiently low temperature, only long wavelength lattice modes are
thermally excited at the centers of interest here, since they do not
show any low-energy resonances. It may be assumed that the quadratic coupling to two of these modes is proportional to the product
of their linear coupling (Eq. 13.40 of [195]). The zero-phonon
broadening function then has a full width at half height of (Eq.
13.44 of [195]):

$$W(T) = d \int d\omega \, [g(\omega)]^2 n(\omega)[n(\omega) + 1] \tag{30a}$$

There is one new undetermined parameter d, but at least the equation
expresses W(T) in terms of the known function $g(\omega)$ rather than a
new quadratic coupling function. The observed zero-phonon line
will be a convolution of this Lorentzian broadening with the zero-
phonon lineshape caused by the completely independent inhomogeneous
broadening in the imperfect diamonds (Eq. 45, Sec. VII). The
properties of this convolution can be evaluated analytically [201].
To a good approximation, the total bandshape is the convolution of
the shape calculated using the linear coupling theory of Eqs. (24)
through (28) with the shape of the zero-phonon line observed at
the same temperature.

Figures 19 through 21 show fits to the 2.985-eV(N3),
3.150-eV(ND1), 3.188-ev, 2.086-eV, and 5.26-eV(N9) bands calculated
with the $g(\omega)$ of Fig. 22 and with S measured for each band from the
ratio of the zero-phonon line to the total band intensity at low
temperature. The zero-phonon intensity may be calculated explicitly
for each band from its low-temperature S-value using $g(\omega)$ [Eq. (29)].
Excellent agreement is found with experiment (Fig. 23).

According to Eq. (30), we should now be able to fit the temperature dependence of the zero-phonon linewidth with just one adjustable
parameter d, as long as the temperature is sufficiently low to jus-
tify the use of $g(\omega)$. In practice, the fits are poor (Fig. 24).
Studying the integrand in Eq. (30a), we see that there are large
contributions from short wavelength acoustic modes at temperatures
as low as 100 K: Fig. 25 gives these data specifically for the N3
band. It appears that the quadratic coupling must decrease with

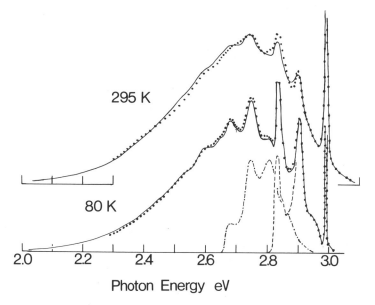

FIG. 19. 2.985-eV(N3) luminescence spectrum. The points give
the measured spectrum in number of photons per unit energy range.
The solid line is the calculated fit assuming coupling to totally
symmetric phonons only. For the 80 K curve the one- and two-phonon
contributions to the luminescence are given explicitly by the broken
and chain lines. The g(ω) used in the calculation is given in Fig. 22.

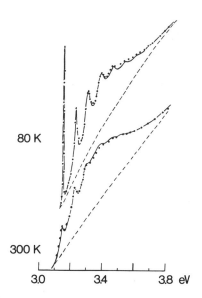

FIG. 20. Measured absorption spectrum (points) and calculated
curve (solid line) for the 3.150-eV(ND1) absorption band. The base-
lines (broken lines) are consistent with other experimental spectra
and with the calculation. The g(ω) is given in Fig. 22. See also
Fig. 44.

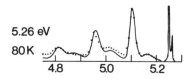

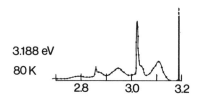

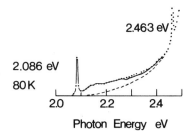

Photon Energy eV

FIG. 21. Experimental spectra (points) and calculated fits (solid lines) for the 5.26-eV(N9), 3.188-eV, and 2.086-eV bands at 80 K. Data for 5.26-eV band from Wight and Dean [147]. The base-line to the 2.086-eV band was calculated to give a self-consistent fit to the calculation; the baseline determines S, which is used in the calculation to find the bandshape. The calculated bandshape must agree with the experimental shape determined using the same baseline. A unique baseline can be chosen since S and the bandshape have differ-ent functional dependences on the position of the baseline [Eq. (28)]. Later experimental work confirmed the baseline; see Fig. A.2.

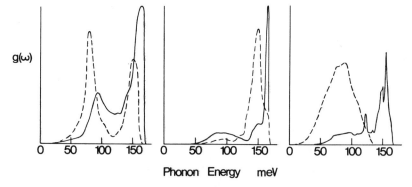

Phonon Energy meV

FIG. 22. The $g(\omega)$ used in the calculations for Figs. 19 through 21. From left to right: 2.985-eV(N3) band (—) and 3.150-eV(ND1) band (---); 3.188-eV band (—), 5.26-eV band (---); 2.086-eV band (---). Solid curve at extreme right gives $\omega p(\omega)$, where $p(\omega)$ is the density of phonon states for perfect diamond calculated by Wehner et al. [21]. The normalization of each $g(\omega)$ is arbitrary.

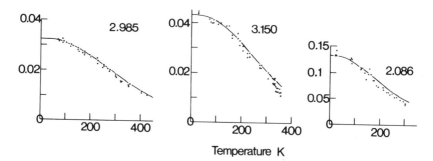

FIG. 23.   Temperature dependence of the 2.985-eV(N3), 3.150-eV
(ND1), and 2.086-eV zero-phonon transition probability as a fraction
of the total band transition probability.   Points are experimental
data, lines are calculated using Eqs. (28) and (29) with the g(ω)
of Fig. 22.   Data for 2.985-eV line taken by Halperin and Nawi
[200].

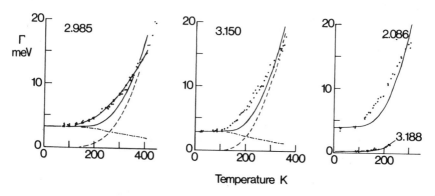

FIG. 24.   Temperature dependence of the full width at half
height of the 2.985-eV(N3), 3.150-eV(ND1), 2.086-eV, and 3.188-eV
zero-phonon lines.   Points are experimental data.   Solid lines are
least-squares fits of Eq. (30) to the data using the g(ω) of
Fig. 22.   The broken lines give the phonon contribution.   The inhomogeneous broadening is assumed to be independent of temperature
so that the additive contribution to the width decreases slowly, as
given by the chain lines.   The line through the points for the
2.985-eV line is calculated using a Debye spectrum, $g(\omega) \propto \omega^3$, with a
Debye cut-off of 60 meV.   Data for 2.985-eV line taken by Halperin
and Nawi [200].

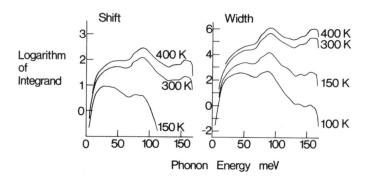

FIG. 25.   Integrands in Eqs. (30a) and (30b) for the temperature
dependence of the peak energy and full width of zero-phonon lines
using the g(ω) for the 2.985-eV band. Note how the width integral
is determined essentially by short wavelength acoustic modes at temperatures as low as 150 K. The width expression, Eq. (30a), is
therefore not necessarily valid at these temperatures. the normalization of g(ω) is arbitrary; the ordinate is a logarithmic scale.

increasing phonon energy [201]. It is worth noting that the linewidths can be fitted exactly, if the Debye approximation is used for
g(ω) and if the Debye cut-off is adjusted for the best fit. This is
illustrated for the 2.985-eV line in Fig. 24, the Debye cut-off
there is 60 meV, which is unreasonably low. Evidently, it is possible to parameterize these thermal data using the Debye approximation, even though the resulting parameters may have no physical significance.

Another effect of quadratic coupling is to shift the first moment of a spectrum as the temperature increases. The movement is
usually to lower energy. Again this effect is most readily seen at
the zero-phonon line. Using g(ω) to represent the quadratic coupling,
the shift is (Eq. 13.43 of [195]):

$$E(T) = f \int d\omega \ g(\omega) n(\omega) \qquad (30b)$$

Inspection of this integrand (shown in Fig. 25 for the 2.985-eV
line) shows that the long wavelength modes are dominant over most
of the useful temperature range, and so the expression should be

valid. The total shift of the zero-phonon line is then the sum of
E(T) and a lattice expansion shift L(T):

$$L(T) = -A(c_{11} + 2c_{12}) \int_0^T e(T) \, dT \qquad (31)$$

where

$$e(T) = \text{coefficient of volume expansion [213]}$$
$$c_{11} \text{ and } c_{12} = \text{compliance coefficients for perfect diamond [214]}$$
$$A = \text{shift of the line under a unit hydrostatic compressional strain}$$

All these are measurable so that f is the only unknown in the shift.
Close fits to the experimental points result (Fig. 26). The
2.086-eV line is unusual in having no detectable shift of its zero-
phonon line from 20 to 300 K. The line has a particularly large
hydrostatic stress shift, $A = 0.65 \times 10^{-8}$ meV Pa^{-1} [203], and this
apparently cancels the E(T) phonon shift. (For comparison the

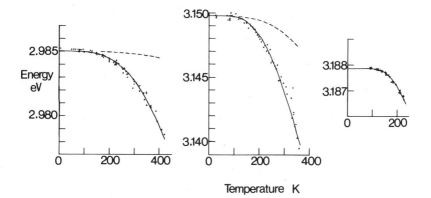

FIG. 26. Temperature dependence of the 2.985-eV(N3), 3.150-eV
(ND1), and 3.188-eV zero-phonon peak energies. Points are experimental data, solid lines are total fits, being the sum of a lattice
expansion term, Eq. (31), which is shown by the broken line, and a
quadratic coupling term which is calculable from the $g(\omega)$ in Fig. 22.
The hydrostatic stress coefficient A [Eq. (31)] is 0.051×10^{-8} meV Pa^{-1}
for the 2.985-eV line, 0.365×10^{-8} meV Pa^{-1} for the 3.150-eV line,
and is unknown for the 3.188-eV line. Data for 2.985-eV line
taken by Halperin and Nawi [200].

3.150-eV (ND1) line has $A = 0.356 \times 10^{-8}$ meV Pa^{-1} [204].) The
E(T) shift of the 2.086-eV line is also unusual in being to higher
energy as the temperature increases.

The main use of these calculations at the moment is that they
establish the simplicity of these vibronic bands. In particular,
there are no complications from Jahn-Teller effects even though the
2.985-eV (N3) and 3.150-eV (ND1) bands are known to involve degenerate excited states. The absence of Jahn-Teller coupling at these
centers is confirmed by uniaxial stress measurements, as we shall
see in the next two sections.

B. The 3.150-eV Band

The 3.15-eV (ND1) line is only observed after radiation damage has
been induced in a diamond containing nitrogen in some form [48,
204-207]. The zero-phonon line responds to uniaxial stresses in a
way characteristic of an electric dipole transition at a tetrahedral
(T_d) center. The ground state has the A_1 (or A_2) irreducible representation and the excited state T_2 (T_1) [204]. The center is more
readily created in type Ib than type Ia diamond [206]. All these
results are consistent with the 3.150-eV center being an interstitial nitrogen atom created by a carbon interstitial atom returning to a lattice site by displacing the nitrogen atom [204]. Similar
processes occur in the other Group IV semiconductors [208].

An A to T transition at a T_d center is affected by perturbations
of A_1, E, and T_2 symmetry only. The perturbation of the Hamiltonian
when stresses are applied is then best written in linear combinations
of the stress tensor components s_{ij}, which transform as A_1, E, and T_2:

$$H' = c_A(s_{xx} + s_{yy} + s_{zz}) + c_{E\theta}(2s_{zz} - s_{xx} - s_{yy})$$

$$+ c_{E\epsilon} (3)^{\frac{1}{2}}(s_{xx} - s_{yy})$$

$$+ c_{yz}s_{yz} + c_{zx}s_{zx} + c_{xy}s_{xy} \tag{32}$$

Using the A_1 to T_2 transition with T_2 having the bases $|Tx\rangle$, $|Ty\rangle$,
$|Tz\rangle$, we define

$$A_T = <Tx|c_A|Tx>, \quad B = \frac{2<Tx|c_{E\epsilon}|Tx>}{(3)^{\frac{1}{2}}}$$

$$C = <Tx|c_{xy}|Ty>$$

and, for the ground state,

$$A_A = <A|c_A|A>$$

Solution of the secular determinant for stresses along the major
axes and comparison with experiment gives the stress parameters
$A = (A_T - A_A)$, B, and C [204]

Vibronic properties are concerned with atomic movements (i.e.,
strains) rather than with applied stresses. The stress Hamiltonian,
Eq. (32), can be written in the same form but with s_{ij} replaced by
the strain e_{ij}, the strain parameters then being

$$a = (c_{11} + 2c_{12})A = 4.72 \text{ eV per unit strain}$$

$$b = (c_{11} - c_{12})B = 1.56 \text{ eV per unit strain}$$

$$c = c_{44}C = 2.26 \text{ eV per unit strain} \tag{33}$$

the c_{ij} are those of McSkimmin and Andreatch [214]; they apply to
perfect diamond, and we must assume they also apply to local strains
near the 3.150-eV (ND1) center. It is pleasing that the strain parameters are of the order of magnitude expected in the simple box model,
Eq. (23).

Now to understand the vibronic properties, we need the strain
coupling to electronic states, not to the zero-phonon line. The
electronic parameters a_e, b_e, c_e may be larger than a, b, c if there
is Jahn-Teller coupling at the center; this is Ham quenching [209,
210]. For an A to T transition, we can select the following special
cases [210]:

1. For coupling to A_1 modes only,

$$a = a_e, \quad b = b_e, \quad c = c_e \tag{34a}$$

2. For coupling only to one pair of E modes of vibrational
    frequency ω and with a Jahn-Teller energy ξ,

$$a = a_e, \quad b = b_e, \quad c = c_e \exp\left(\frac{-3\xi}{2\hbar\omega}\right) \tag{34b}$$

3. For coupling only to one trio of T_2 modes,

$$a = a_e, \quad b = b_e \exp\left(\frac{-9\xi}{4\hbar\omega}\right)$$

$$c = c_e \left[\frac{2 + \exp(-9\xi/4\hbar\omega)}{3}\right] \tag{34c}$$

4. For equal coupling to one pair of E modes and one trio of
    T_2 modes, O'Brien [211] has established that

$$a = a_e, \quad b \geq 0.4b_e, \quad c \geq 0.4c_e \tag{34d}$$

In this case the quenching is never large.

At the 3.150-eV center, the total energy reduction ξ through
the electron-lattice coupling is, from Fig. 23,

$$\frac{\xi}{\hbar\omega} \sim -\ln I_0 = 3.15$$

where I_0 is the normalized zero-phonon transition probability. Very
strong quenching would occur for parameters c or b, if this coupling
was purely to E or to T_2 modes in contrast with the reasonable size
observed for these parameters [Eq. (33)]. We can therefore exclude
cases (2) and (3). Case (1) is more acceptable, except that (b)
and (c) as measured are smaller than (a) but not really negligible in
comparison with it. It seems likely, therefore, that there is roughly
equal weak coupling to E and T_2 modes, case (4). However, the dominant coupling is to A_1 modes, as we can see through a cluster model
approach.

The T_d interstitial site has four nearest neighbors at a distance of 0.154 nm in the undistorted lattice. There are six next
nearest neighbors distant 0.181 nm. The 3.150-eV center can be represented, then, by n = 10 neighbors at essentially the same separation, say R $\sim$ 0.181 nm from the nitrogen atom. The Huang-Rhys factor
for coupling to an A_1 mode of frequency ω is readily found to be

$$S_A = \frac{9a_e^2}{2nR^2Mh\omega^3} \tag{35}$$

where M is the atomic mass of carbon. Evaluation gives S_A = 2.3
with the 3.150-eV parameters a = 4.72 eV, n = 10, r = 0.181 nm,
$h\omega$ = 78 meV. This compares well with the experimental value of
3.15 from Fig. 23. A similar analysis for the E and T_2 mode coupling
requires a knowledge of how many atoms are moving [cf. n in Eq. (35)].
An upper limit to the coupling will come from using four neighboring
atoms for which [210, 212]

$$S_E = \frac{9b_e^2}{4R^2Mh\omega^3}$$

and

$$S_T = \frac{c_e^2}{4R^2Mh\omega^3} \tag{36}$$

but more likely values are about half these since n $\sim$ 10. Evaluation
with the quenched values b and c gives S_E = 0.6 and S_T = 0.1. These
values could be about twice as large allowing for the Jahn-Teller
effects, case (4).

All three values, S_A, S_E, and S_T, are uncertain through such
unknowns as the local elastic properties. But the cluster model
calculations verify the conclusion of Sec. V.A that Jahn-Teller
effects are unimportant at the 3.150-eV center.

C. The 2.985-eV Band

The 2.985-eV (N3) band is observed frequently in natural type Ia diamonds, especially in those with large B nitrogen concentrations. It
is probably the most studied band in diamond. It was first reported
in 1891 by Walter [215] and was perhaps the first band in any crystal
to be recognized as having a vibronic origin [216].

Uniaxial stress [202, 217], Stark effect [218], and polarized
luminescence [219] experiments have established that the 2.985-eV

line is an electric dipole transition between A_1 ground and E excited
states of a C_{3v} optical center. The zero-phonon line is a doublet of
0.59 ± 0.02 meV [47]. The splitting is in the E state, presumably
a spin-orbit interaction [220]. An electron spin resonance which has
been identified as due to a spin one-half at a triangle of nitrogen
atoms [221, 222] has been tentatively correlated with the 2.985 eV
center [221]. A working model for the center is thus a triangular
arrangement of substitutional nitrogen atoms (as guessed in the early
days by Mitchell [223]). Each nitrogen atom is bonded to one, common
carbon atom, the four atoms forming an almost planar complex in the
distorted lattice. The plane lies perpendicular to a $< 1\ 1\ 1 >$ trigonal axis.

To discuss the uniaxial stress data, we concentrate on the
[1 1 1]-oriented center. Stress terms of A_1 and E symmetry need
to be considered. The perturbation Hamiltonian becomes

$$H' = c_A(s_{xx} + s_{yy} + s_{zz}) + c_A'(s_{yz} + s_{zx} + s_{xy})$$

$$+ c_{Ex}(2s_{zz} - s_{xx} - s_{yy}) + c_{Ey}(3)^{\frac{1}{2}}(s_{xx} - s_{yy})$$

$$+ c_{Ex}'(2s_{xy} - s_{yz} - s_{zx}) + c_{Ey}'(3)^{\frac{1}{2}}(s_{yz} - s_{zx}) \qquad (37)$$

Using the notation of Hughes and Runciman [224], we define

$$A_1 = <Ex|c_A|Ex>, \quad A_2 = \tfrac{1}{2}<Ex|c_A'|Ex>$$

$$B = \frac{<Ex|c_{Ex}|Ex>}{(2)^{\frac{1}{2}}}, \quad C = \frac{<Ex|c_{Ex}'|Ex>}{(2)^{\frac{1}{2}}}$$

These coefficients can be calculated from the stress data of Crowther
and Dean [202]. For vibronic properties, we are more interested
in strain effects and so replace s_{ij} by the strain component e_{ij} of
the same symmetry. The corresponding strain coefficients are

$$a_1 = (c_{11} + 2c_{12})A_1 = 0.70 \text{ eV per unit strain}$$

$$a_2 = c_{44}A_2 = 2.47 \text{ eV per unit strain}$$

$$b = (c_{11} - c_{12})B = -1.05 \text{ eV per unit strain}$$

$$c = c_{44}C = -0.85 \text{ eV per unit strain}$$

Now these coefficients refer to the crystallographic axes x, y, z. When discussing the properties of a [1 1 1]-oriented center, it is preferable to transform into Cartesian coordinates, X, Y, Z defined with respect to the center where Z is the [1 1 1] trigonal axis, X is [1 1 $\bar{2}$], and Y is [$\bar{1}$ 1 0]. The X-axis lies in one of the reflection planes of C_{3v}. The strain Hamiltonian is now

$$H' = P(e_{XX} + e_{YY} + e_{ZZ}) + Q(2e_{ZZ} - e_{XX} - e_{YY})$$

$$+ R_X(e_{YY} - e_{XX}) + 2R_Ye_{XY} + 2S_Xe_{YZ} - 2S_Ye_{XZ} \tag{38}$$

We define new strain coefficients for these local coordinates:

$$p = \langle EX|P|EX\rangle = a_1 = 0.70 \text{ eV}$$

$$q = \langle EX|Q|EX\rangle = a_2 = 2.47 \text{ eV}$$

$$r = \frac{\langle EX|R_X|EX\rangle}{(2)^{\frac{1}{2}}} = b + c = -1.90 \text{ eV}$$

$$s = \frac{\langle EX|S_X|EX\rangle}{(2)^{\frac{1}{2}}} = \frac{2b - c}{(2)^{\frac{1}{2}}} = -0.88 \text{ eV}$$

The uncertainties are about ±0.15 eV.

As far as the vibronic properties are concerned, the significant point is that q is dominant, and q measures the response to dilationless symmetry maintaining vibrations along the trigonal axis. One such mode is the Raman mode, and we do see high-energy optical modes in $g(\omega)$, Fig. 22.

VI. NONSIMPLE VIBRONIC SPECTRA

A. The 1.673-eV Band

In Sec. V, we saw that some of the optical bands in diamond are explicable in terms of the standard theory of total symmetric electron-lattice coupling. We now turn briefly to some of the "nonsimple" centers where these ideas do not apply.

It is immediately obvious from Fig. 27 that the 1.673-eV (GR1) band does not have totally symmetric coupling. For example, there is no combination band formed from the resonance at 1.71 eV and the optical modes at 1.838 eV. The combination would lie in the local minimum of absorption near 1.88 eV. To date, no satisfactory explanation of the known properties of the band has been presented and in fact the experimental picture is not yet complete.

The 1.673-eV center is produced in any type of diamond when radiation damage is created [48]. In natural type Ib diamond the absorption is weaker than the same radiation dose would produce in other diamonds, apparently as a result of charge transfer effects

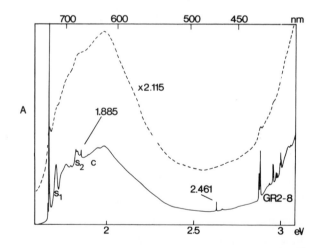

FIG. 27. Absorption of a type IIa diamond at 80 K (solid line) and 300 K (broken line) after irradiation with 2-MeV electrons. Assuming the peaks s_1, s_2 are one-phonon sidebands of the 1.673-eV line, note that there is no combination mode; it would occur at c. See also Figs. 28 and 29.

(Sec. VIII). A report that the 1.673-eV center was not present in
type IIb diamond until all the acceptor centers were compensated
[225] has recently been shown to be wrong [226]. Consequently, the
conclusion that the 1.673-eV center is a donor (and so is a negative
or neutral vacancy) is also suspect. What is certain is that the
center does not involve impurities. Since it is stable to very high
temperatures (1200 K), we can rule out the carbon interstitial atom
and so are left with the center being a vacancy. Certainly it has a
T_d symmetry as we see in the following discussion.

At the same time as the 1.673-eV band is produced, a series of
absorption lines appear in the blue part of the spectrum (Fig. 28),
and a long continuum stretches to the absorption edge [26]. The
sharp structure is caused by transitions from the 1.673-eV ground

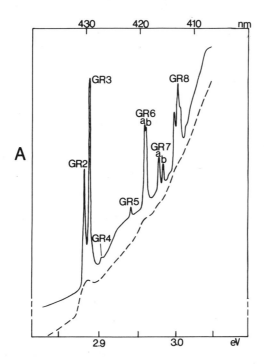

FIG. 28. Features associated with the 1.673-eV center on the
ultraviolet continuum of absorption in irradiated diamond. Solid
line at 80 K, broken line at 300 K. See also Fig. 29.

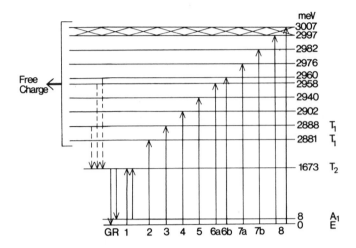

FIG. 29. Energy level diagram for the sharp absorption lines of the 1.673-eV (GR1) center. Vertical upward arrows show the observed absorption transitions. The lines are labeled along the bottom and their energies at the extreme right, as are the irreducible representations, where known, of the levels. (All the subscripts are ambiguous: 1 and 2 can be interchanged.) Vertical downward arrows with solid lines show the observed luminescence lines. Broken downward arrows indicate the de-excitation paths which must exist to be consistent with luminescence excitation work [229]. Horizontal arrow at left shows those states which emit a free charge [271]. GR8 is a multiplet between extrema of 2.997 and 3.007 eV. See also Figs. 27 and 28.

state to higher internal levels of the center (Fig. 29) [227]. Absorption into the ultraviolet continuum produces free positive holes [228] and weakens the 1.673-eV intensity (Sec. VIII). The liberation of holes is compatible with the 1.673-eV center being able to co-exist with neutral acceptor centers in semiconducting type IIb diamond. Surprisingly excitation into the ultraviolet continuum also produces 1.673-eV luminiscence [229]. Some of the energy must remain localized on the 1.673-eV centers, in contrast to the break-up of the center, implied by the photoconduction. The time taken to emit a hole must therefore be of the order of the time required for the center to de-excite to the 1.673-eV level (see Fig. 29).

The 1.673-eV zero-phonon line is in fact a doublet [230]. The two components are at 1.665 and 1.673 eV at 80 K. Their integrand absorption intensities follow [230]

$$\frac{I(1.665 \text{ eV})}{I(1.673 \text{ ev})} = \frac{m_2}{m_1} \exp\frac{-E}{kT}, \quad E = 8 \text{ meV}, \quad 10 < T < 130 \text{ K} \quad (39)$$

The ratio $m_2/m_1 = 0.155 \pm 0.025$. There is no Debye–Waller factor in the intensity ratio, since this term is essentially constant over the temperature range used in comparison with the Boltzmann term.

The Boltzmann factor in Eq. (39) shows the doublet splitting to be in the ground state. Luminescence of the 1.673-eV center is not easy to detect since its decay time is only 3 nsec [232]. Electric dipole allowed transitions like the 1.67-eV lines [233, 234] have radiative lifetimes about ten times longer than this (Appendix Tables A1 and A2). There is, probably, an internal nonradiative decay path in competition with the radiative transition. However, the luminescence spectra available confirm that there is a large intensity ratio between the 1.665- and 1.673-eV lines as required by Eq. (39) [235, 236]. Inspection of the total intensity of the absorption lines as a function of temperature enables limits to be placed on the degeneracies and hence irreducible representations of their ground states. At a vacancy (T_d symmetry), the two zero-phonon lines are found to be from an A to a T state for the 1.665-eV line and from E to T for the 1.673-eV line [231]. The response to uniaxial stresses is consistent with this and further limits us to:

1.665 eV:

$$A_1 \text{ to } T_2 \quad A_2 \text{ to } T_1$$

or

1.673 eV:

$$E \text{ to } T_2 \quad E \text{ to } T_1$$

The perturbation produced by the uniaxial stresses is as Eq. (32). It mixes the A and E ground states, producing nonlinear shift rates of the zero-phonon lines. We define stress coefficients

$$A_E = \langle\theta|c_A|\theta\rangle, \quad D = \langle\theta|c_{E\varepsilon}|\varepsilon\rangle, \quad E = \langle A|c_{E\varepsilon}|\varepsilon\rangle$$

for the matrix elements of the operators c_A, $C_{E\epsilon}$ in Eq. (32) between
the θ, ϵ components of E and the A ground states. Other necessary
coefficients are defined in Sec. V.B. The fit to the experimental
shifts (Fig. 30) gives these stress coefficients [234]. Transforming
to strains we have:

$$a_T - a_A = (A_T - A_A)(c_{11} + 2c_{12}) = -325 \pm 400 \text{ meV per unit strain}$$

$$a_T - a_E = (A_T - A_E)(c_{11} + 2c_{12}) = 425 \pm 400 \text{ meV per unit strain}$$

$$b = B(c_{11} - c_{12}) = -790 \pm 300 \text{ meV per unit strain}$$

$$c = C\, c_{44} = 2745 \pm 200 \text{ meV per unit strain}$$

$$d = D(c_{11} - c_{12}) = -2375 \pm 300 \text{ meV per unit strain}$$

$$e = E(c_{11} - c_{12}) = \pm 3010 \pm 300 \text{ meV per unit strain} \quad (40)$$

The sign of e, which mixes the A and E ground states, is not determined from shift rate measurements. The sign of d is correct for
A_1 to T_2 transitions and needs reversing for A_2 to T_1.

We are now in a position to understand the one major disagreement between the published stress results on the 1.67-eV lines.
Under < 1 1 1 > stress, an E to T transition, like the 1.673-eV
line, should remain as a single line in the spectrum recorded with
electric vector parallel to the stress axis. Only one set of results
shows this [234]. Others show the transition splitting into two
well-resolved lines [e.g., 217, 233]. Now the data with the anomalous splitting were taken with the stress applied to polished
< 1 1 1 > faces. But the < 1 1 1 > surface of diamond is very re-
sistive to abrasion, the resistance decreasing rapidly as one pol-
ishes at an increasing angle to the < 1 1 1 > plane. In practice,
a polished surface which is nominally < 1 1 1 > will be a few degrees
off axis. Often this misorientation is unimportant, but the strain
coefficients in Eq. (40) show that misorientations are extremely
important for stress experiments on the 1.67-eV lines. From Eq. (40),
we see that the largest coefficients are c, d, and e. For brevity,

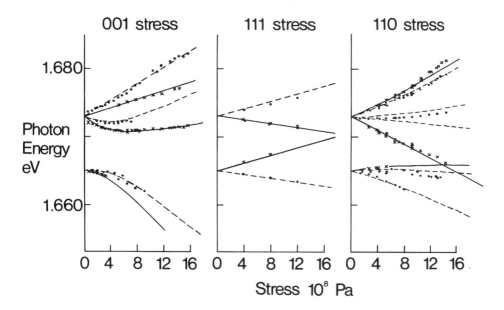

FIG. 30. The effect of uniaxial stress on the 1.673- and
1.665-eV lines is to produce nonlinear shift rates as the A and E
ground states interact. The eighteen experiment shift rates (x for
π polarization, ● for σ) can be fitted using the six coefficients in
Eq. (40), so that there is confidence that these coefficients are
essentially correct. The solid lines are the fits for π polarization, the broken lines for σ polarization. Some lines are recorded
only over limited ranges of stress because of thermal depopulation
of their ground states [234].

we concentrate on d and e which control the interactions of the E

and A ground states with the stress combinations $s' = (2s_{zz} - s_{xx} - s_{yy})$

and $s'' = 3(s_{xx} - s_{yy})$ according to Eq. (32). With the stress

applied along < 1 1 1 >, $s' = s'' = 0$. But if a stress s is applied

at only 5° from [1 1 1], say along [1, 1, 1.198], we find that

$s' = 0.247s$ and it is now of a similar magnitude to each stress

tensor component since $s_{ij} = s/3$ for all i,j under [1 1 1] stress.

The sensitivity of s' to small misorientations and the large values

of d and e produce stress splitting patterns which rapidly deviate

from the < 1 1 1 > case as the stress axis changes. In fact, the

large discrepancies between theory and those experiments performed

with polished < 1 1 1 > faces can be understood quantitatively with
misorientations of 5° [234b].

We have discussed the effects of temperature and stress on the
zero-phonon lines at some length to emphasize that they are fully
understandable in general terms. The problems arise when we try to
understand the strain-coupling parameters. As long as there are no
dramatic changes in the elastic constants near the 1.673-eV center
[237], the parameters show that the coupling to totally symmetric
modes is far smaller than the coupling to Jahn-Teller active modes.
This means that we are right in a Jahn-Teller situation.

We take the excited T-state first. The strain-coupling parameters b and c are both large, predicting, from Eq. (32), that coupling to both "tetragonal" and trigonal modes will occur. The Jahn-
Teller energies [Eq. (36)] in units of the phonon energy are S_E = 0.28
S_T = 0.38. Here we have used the quenched (zero-phonon) strain parameters since the quenching will be small [Eq. (34)]. Phonon energies
of 90 meV have been assumed since these are important in the absorption spectrum [238]. We can picture the vibrational potential in
the electronic excited state as having a shallow circular trough in
the five-dimensional vibrational space of the E- and T_2-modes. The
minimum of this trough is about $0.3\hbar\omega \sim 30$ meV below the electronic
energy. This Jahn-Teller energy is too small to account for the
strength of the phonon sideband, and we must turn to the ground state.
The large value of d indicates stronger coupling to the E modes. In
four-fold coordination the Jahn-Teller energy is given by the same
expression as for S_E in Eq. (36) (Eqs. 6, 39, 40, 43c of [209]).
This gives S_E = 2.56, again using the quenched (zero-phonon) value
for d. For this strength of coupling d will be quenched to one-half
the electronic value [210], i.e., the Jahn-Teller energy is S_E = 10.2.
In this model the zero-phonon line would have an intensity about
$\exp(-S_E)$ of the total band intensity; the experimental value is
about $\exp(-3.5)$, in reasonable agreement. We can picture the vibrational state as having a circular trough in the E-mode space,
the trough being displaced a distance $\rho = 6d/2 \, (2)^{\frac{1}{2}} R M \omega^2 = 1.41 \times 10^{-11}$ m
from the origin of the E-mode space [210]. This relaxation will quench

matrix elements between the E ground state of the 1.673-eV transition and the ground vibronic state of an A electronic level, since the A-state cannot couple to E-modes and so is located near the origin of the E-mode space. The quenching will be by a factor of the order of $\exp(-S_E/2)$ by analogy with zero-phonon intensities. If the coupling of the A and E ground states of the 1.665- and 1.673-eV lines was quenched to this extent, the coupling of their A and E electronic states would be very large ($\sim$15 eV per unit strain). In fact, the measured value of e is already of the magnitude expected for electronic coupling [cf., c in Eq. (23)]. It looks, therefore, as though the A and E ground states of the 1.67-eV doublet are derived from the same electronic state.

The Jahn-Teller effect provides a partial explanation of the E and A ground states [239]. When an E electronic state couples to an E-mode the ground vibronic state transforms as E. At an energy $\xi \sim \hbar\omega/2S_E$ [210] above the ground state, there is a pair of accidently degenerate A-states (A_1 and A_2). Using, again, $\hbar\omega$ = 90 meV and S_E = 3.5, the energy is 13 meV. The degeneracy of A_1 and A_2 is lifted by coupling of a higher order than the linear coupling considered here [240]. A single A state could, presumably, be brought to 7 meV from the E ground state, as observed. Unfortunately the transition from this A state to the T excited state would be forbidden [241]. Even if it is made allowed by some process, it is not clear why the 1.665-eV line should have a width equal to the genuine zero-phonon line at 1.673 eV; the 1.665-eV line would suffer some broadening from the spread of phonon energies and the different strength of coupling to each mode. However, this scheme gets over the problem of the large value of the A to E coupling term e. In fact theoretically, the ratio $e/d = -(2)^{\frac{1}{2}}$ [24], and this is close to the observed value for $|e|/d$, Eq. (40) [243].

One additional problem with this whole scheme is that the absorption spectrum should be quite simple, since there is little Jahn-Teller coupling in the excited electronic state. (An example of this simplification is seen in the calculated E to A spectrum of Longuet-Higgins et al. [244]; see also the 2.463-eV emission in the

next section.) But, as we noted at first, the 1.673-eV absorption
band is not simple: There is no combination band of the 40 and 165
meV phonon sidebands [245].

Clearly there are processes occurring at the 1.673-eV center
that are not understood at present. It is some relief that the
luminescence of the center is not a mirror image of the absorption
spectrum [246] and that the photoluminescence is unpolarized for
all polarizations of the exciting light [229].

On the experimental side, we have no information of the Ham
quenching nor of the central moments of the band. The radiative
lifetime has not been measured as a function of temperature so that
we do not yet know why it is only 3 nsec at 80 K. Certainly there
seems to be no free charge emission associated with the 1.673-eV
absorption [247]. On the theoretical side, one fundamental problem
is that the charge state of the center is unknown. Lanoo and
Stoneham [248] predicted a neutral vacancy would behave somewhat
as observed except that they concentrated on static Jahn-Teller
effects. A second, more basic difficulty with all theoretical calculations is that at the moment there is no known method of predicting the behavior of a vacancy which one can trust to be reasonably reliable. Recent calculations are given in [249].

B. The 1.946-eV Band

When radiation damage in type Ib diamond is annealed, say at 900 K,
the diamond turns a striking mauve color as the 1.946-eV absorption
band is created (Fig. 31). The band can be observed in type Ia diamonds which have been given the same treatment, but the absorption
strength is very much weaker. Therefore it is apparent that the
center involves a single nitrogen atom plus a radiation damage center. There are two reasonable candidates for the damage center, a
vacancy or an interstitial nitrogen atom (the 3.150-eV (ND1) center:
Sec. V.B). Annealing experiments show that the 1.673-eV center, a
vacancy in some charge state, and the 3.150-eV center anneal out at
very similar rates, the 1.946-eV band growing in sympathy.

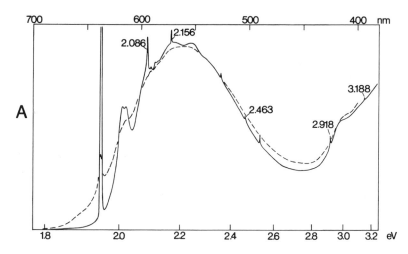

FIG. 31. Absorption spectrum of the 1.945-eV band at 80 K
(solid line) and 300 K (broken line) in an irradiated and annealed
type Ib diamond. There is some sharp structure on the band from
other zero-phonon lines, as shown. Note the doublet nature of the
one-phonon sideband near 2.0 eV: see Fig. 32.

Consequently, it has not yet been established which of the two damage
centers is involved. We give reasons below for favoring the vacancy.

The 1.946-eV center has a trigonal symmetry, as expected from
one nitrogen atom plus one damage center [229, 250]. Uniaxial
stress and polarized luminescence measurements show the zero-phonon
line is a transition between A_1- and E-states, the E being the
excited state [229, 250]. An E-state suggests the possibility of a
Jahn-Teller effect. However, the coupling has been shown to be
essential to totally symmetric modes by studying the effect of
stress on the luminescence spectrum [250]. In luminescence, the
zero-phonon line is an E- to A_1-transition, and so the one-phonon
sideband is E to A_1 or E to E depending on whether A_1- or E-modes
of vibration are important. These transitions are distinguishable
by their different polarization selection rules under stress: The
method is that of Fetterman and Fitchen [251].

Large quadratic electron-lattice coupling at the 1.946-eV
center can be ruled out since the zero-phonon line has a temperature

dependence which is very similar to that of the 2.985-eV line [201].
The absence of Jahn-Teller coupling then implies that the absorption
and emission spectra should be good mirror images of each other.
But, experimentally, the first-phonon sideband is seen as a distinct
doublet in the absorption spectrum, while in the luminescence spectrum it is a single, sharper peak (Fig. 32). The absorption spectrum
generally has less well-defined features than the emission spectrum,
but otherwise, the two are approximate mirror images. The differences, although small, are real effects and occur in all specimens.
Perhaps the major problem is why the one-phonon sideband is qualitatively different in absorption and luminescence.

We can understand the spectra if we assume the 1.946-eV center
is a vacancy-nitrogen pair and not the otherwise equally likely interstitial nitrogen-nitrogen pair. A vacancy-nitrogen pair has the
property that the nitrogen atom can tunnel into the site occupied
by the vacancy, giving an equivalent nitrogen-vacancy pair. This
tunneling will give an "inversion splitting" of each level in which
tunneling can occur, the splitting being by an energy ΔE which

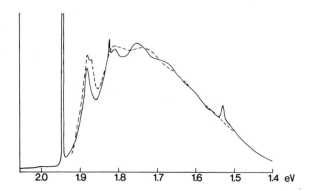

FIG. 32. Photoluminescence spectrum of 1.945-eV band at 80 K.
Solid line is the experimental curve in number of photons per unit
energy range, data by M. F. Hamer. Broken line is the luminescence
expected from the absorption spectrum, Fig. 31, if the absorption
and emission spectra have mirror symmetry. Note the overall simi-
larity of the curves, but the qualitative differences in the one-
phonon sideband near 1.88 eV.

decreases as the time t required to tunnel increases: $\Delta E \sim h/t$.
Perhaps the most familiar example of this type of effect is the inversion line of ammonia.

In the nomenclature of Eq. (38), the strain parameters are [250]:

p = 1.95 eV per unit strain

q = -2.22 eV per unit strain

r = -1.96 eV per unit strain

s = -0.71 eV per unit strain

Cluster model calculations suggest that the uniaxial breathing mode
is most important in producing the vibronic spectrum. This motion
is that naturally involved in the tunneling effect. We can use
the splitting of the one-phonon sideband seen in the absorption
spectrum together with the lack of observed splitting in the zero-
phonon line to define the vibrational potential in the excited
electronic state. Using the theoretical expressions derived by
Dennison and Uhlenbeck for their model of ammonia [252] we obtain
the upper vibrational potential curve of Fig. 33. The curvature
of these parabolae is given by the observed, mean phonon energy,
70 meV, and the reduced mass of the oscillating cluster. The

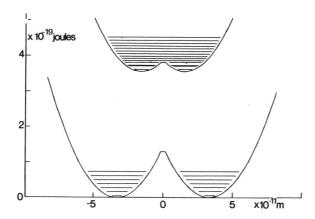

FIG. 33. Vibrational potential energy curves in the ground and
excited electronic states of the 1.946-eV center. The horizontal
lines are the total energy eigenvalues.

eigenvalues for energies greater than the plateau level can be found
by the W.K.B. method [253]. This requires the classical momentum p
of the oscillating system to obey the Wilson-Sommerfeld quantization
of action, the action integral being over that region of space x
allowed to the oscillator in classical mechanics:

$$\oint p \ dx = (n + \tfrac{1}{2})h, \quad n = 4,5,6,\ldots$$

These eigenvalues are on Fig. 33. The vibrational potential in the
electronic ground state is defined by a further pair of parabolae
[254]. Their curvature is the same as in the excited electronic
state, since we have seen we can neglect quadratic electron-lattice
coupling. Their separation is increased relative to the excited
state so that the orthogonality of the two sets of vibrational wavefunctions is broken and vibronic transitions can occur. The change
in position of the minima is determined by the Huang-Rhys factor,
which can be defined as in the simple totally symmetric coupling
theory, Sec. V.A. The change is by 1.77×10^{-11} m (Fig. 33). The
two ground state parabolae are now sufficiently separated that no
tunneling effects can occur. Therefore, we obtain our first result
that the luminescence spectrum will be an orthodox totally symmetric
type of spectrum, as observed. The absorption spectrum is more complicated since transitions are now being made into states which will
be strongly affected by the two minima in the potential. To calculate the absorption spectrum, we need the wavefunctions of the oscillator in the electronic excited state. These can be found by
the W.K.B. method, except for the energy levels immediately above
the plateau. The probabilities of transitions into these states can
then be calculated. The resulting spectrum has a width much as observed (Figs. 31 and 34). There are more transitions per unit energy
than for the luminescence spectrum, so that the absorption features
are the less well defined, again as observed. The features of the
spectra are closely reproduced by this model, and we note that the
dimensions derived for the center are not implausible.

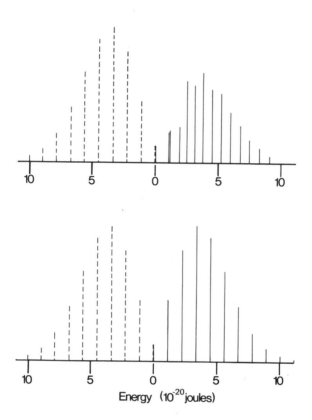

FIG. 34. At top, absorption and luminescence transition probabilities calculated from the eigenstates of the potentials in
Fig. 33. The absorption spectrum is shown by the solid line, the
luminescence by the broken line. The fourth-phonon sideband of the
absorption spectrum is probably anomalously strong because of the
inadequacy of the W.K.B. approximation in dealing with states whose
wavelength is long relative to changes in the potential. The high-
energy transitions should be increasingly accurate, however. At
bottom are the absorption and luminescence spectra for the usual,
single parabolae situation. Note how small are the differences in
the two sets of spectra in spite of the radically different
potentials.

It still remains to prove beyond doubt that the 1.946 eV center
is a nitrogen-vacancy pair. But it is difficult to see how any other
model would explain the small but very real differences in the absorption and luminescence spectra.

C. The 2.463- and 2.499-eV Bands

The 2.463- and 2.499-eV (H3 and H4) bands are two of the most commonly observed optical centers in diamond. They are formed when
radiation damage is annealed in type Ia diamond. A 2.463-eV center
results when a damage center is trapped at an A aggregate of nitrogen;
a 2.499-eV center is formed when the same species of radiation damage
center is trapped at a B aggregate [255]. In any diamond, the ratio
of 2.463- to 2.499-eV strengths is then proportional to the A to B
infrared absorption strengths, apart from any problems of specimen
inhomogeneity. A graph of 2.463/2.499 eV versus $A_A(1282$ cm$^{-1})/$
$A_B(1282$ cm$^{-1})$ has a slope of the order of unity (Fig. 35), since
the 2.463- and 2.499-eV bands are very similar to each other
(Fig. 36).

Annealing experiments [207] suggest that the radiation damage
defect involved in forming the 2.463- and 2.499-eV centers is probably the 3.150-eV (ND1) center [47]. Since the 2.463- and 2.499-eV
centers are intimately connected with the A and B forms of nitrogen,
any information about their structure will be useful in limiting the
number of possible models for the nitrogen impurity. Polarized
luminescence data for the 2.463-eV band are unanimously consistent
with its being a $\pi < 1 1 0 >$ oscillator, that is, an electric dipole
transition which is stimulated by the $< 1 1 0 >$ component of the
electric vector of light, the transition occurring between nondegenerate states [219]. The symmetry classes for the center are then
either rhombic I, i.e., C_{2v} point group with the C_2-axis along
$< 0 0 1 >$ and σ_v in $< 1 1 0 >$, or monoclinic I, e.g., C_{1h} with σ_h
in $< 1 1 0 >$.

Rhombic I and monoclinic I symmetries are distinguishable by
uniaxial stress experiments. Recent work in this laboratory has

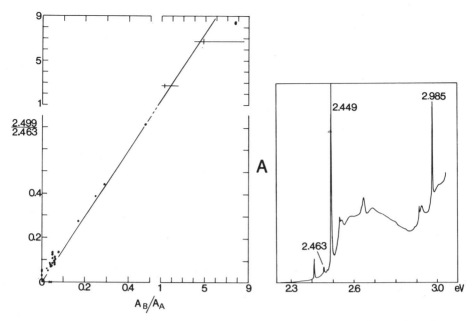

FIG. 35. At left, the ratio of the absorption at the peaks of the 2.499- and 2.463-eV zero-phonon lines compared with the ratio $A_B(1282\ cm^{-1})/(A_A(1282\ cm^{-1})$ in the same specimens. Only some of the specimens are shown for $A_B/A_A < 0.1$. For large values of A_B/A_A, the precise value is difficult to determine, see Fig. 6. At right is shown a diamond with no detectable A absorption relative to the B absorption. The 2.499/2.463-eV peak heights exceed 40:1.

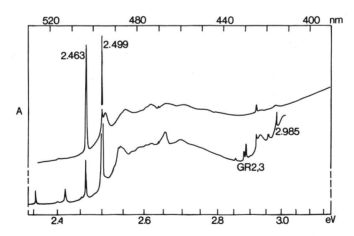

FIG. 36. Comparison of the absorption spectra at 80 K of two electron-irradiated and -annealed type Ia diamonds. Upper curve shows predominantly 2.463-eV band, lower curve predominantly 2.499-eV band. Note how the two bands are broadly similar; the major apparent difference is the relatively strong 154-meV phonon sideband of the 2.499-eV line compared with the 152-meV sideband of the 2.463-eV line. Additional structure near 2.4- and 2.92-eV is often seen accompanying the 2.499-eV line; see also Fig. 35.

shown the 2.463-eV line is an A_1 to B_2 (or B_2) transition at a rhombic I center [256a]. This is in contrast to the early stress experiments of Runciman [217] who stated that the spectra he observed under
< 1 1 1 > stress had too many components to be consistent with rhom-
bic I symmetry. But we have already noted (Sec. VI.A) that Runciman's
< 1 1 1 > stress was misaligned by a few degrees. Using the stress
coupling coefficients derived from Hamer's work [256a], it is
straightforward to show that the strong absorption expected with
electric vector parallel to the < 1 1 1 > stress [258] would be split
into two clearly resolved lines as observed by Runciman by a 5°
stress misorientation. Polarized luminescence [219] and uniaxial
stress experiments [217, 256a] on the 2.463-eV line are therefore
consistent with an A_1 to B_1 (or B_2) transition at a rhombic I C_{2v}
center.

The temperature dependence of the 2.463-eV line is typical of
zero-phonon lines in diamond [201], showing there is not unusually
large quadratic electron-phonon coupling at the center. Since the
lowest vibronic states have been shown to be nondegenerate, we expect the luminescence and absorption spectra to be close mirror
images in the zero-phonon line. This is not the case. The luminescence spectrum is a simple vibronic progression (Fig. 37) while
the absorption has a strong sideband 154 meV above the zero-phonon
line (Figs. 36 and 37). The lack of mirror symmetry is not understood at present.

Jahn-Teller effects have sometimes been invoked to explain the
properties of the 2.463-eV band [203]. An apparently rhombic I
center is produced when a tetragonal center undergoes a large Jahn-
Teller distortion through coupling of an E electronic state to a B_2
mode of vibration. One of the four stress coefficients, B in Hughes
and Runciman's notation [224], is Ham-quenched towards zero by the
distortion, leaving three nonzero coefficients which have the same
effects as the three coefficients of a rhombic I center [257]. A
sufficiently large Jahn-Teller effect to quench would be difficult
to detect, since the coupled B_2 mode is nondegenerate. Two distinguishable potential minima exist in contrast to the one

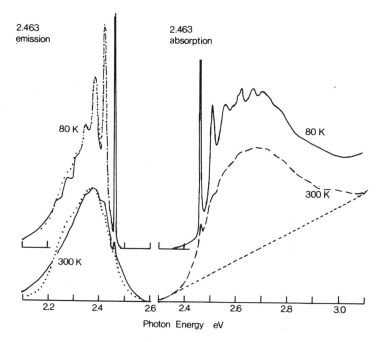

FIG. 37. At left, luminescence of the 2.463-eV band at 80 K and 30 K. Points are experimental data in number of photons per unit energy, lines are calculated using coupling to totally symmetric modes only [201]. At right are absorption spectra of the 2.463-eV band at 80 K and 300 K. The baseline is a guide line only.

trough-like minimum of an E-mode, Jahn-Teller effect. As a result of the insularity of the two potential minima, stress-induced dichroism of the zero-phonon line will not occur even though the initial state of the transition is double degenerate.

Clark and Norris [219] have suggested that a weak absorption line at 3.368 eV is caused by a transition from the ground state of the 2.463-eV center to a higher excited state. There seems to be little evidence in favor of this model at the moment.

Polarized luminescence studies of the 2.499-eV band by Clark and Norris are in closest agreement with a $\pi < 1\ 1\ 0 >$ oscillator [229]. Uniaxial stress measurements have now confirmed this and established the symmetry to be monoclinic I and so of C_{1h} point group [256b]. The stress-coupling coefficients are very similar to the

corresponding ones of the 2.463-eV line with the extra monoclinic I
parameter, A_4 in Kaplyanskii's notation [257], being relatively small.
This raises the same problem as for the 2.463-eV center: The 2.499-eV
luminescence band is a simple progression [258], very similar to the
2.463-eV luminiscence of Fig. 37, in contrast to the more complicated
absorption band, Figs. 35 and 36.

An absorption band with zero-phonon line at 2.417 eV is usually
observed when the 2.499-eV band is present (Figs. 35 and 36). This
line is also a transition at a monoclinic I center, and the stress-
coupling coefficients are similar to those of the 2.463- and 2.499-eV
lines [256b]. Presumably, all three centers are very similar, raising
the possibility of a third type of nitrogen, the "C form," in type Ia
diamonds. Since the 2.417-eV line is typically a factor of 10 weaker
than the 2.499-eV line, this C form would also be expected to be
about an order of magnitude less abundant than the B form, by analogy
with the left diagram in Fig. 35. Relative concentrations of this
order would not be readily detectable in the impurity induced infrared absorption (Sec. III.B). Certainly if the C form exists, it is
reasonable to expect it to appear in type IaB diamonds as already
there is evidence that they probably contain nitrogen in minor
forms such as the 2.985-eV (N3) center (Sec. V.C).

VII. PERTURBATIONS OF OPTICAL CENTERS BY NITROGEN

In the last few years, it has become clear that the nitrogen impurity
in diamond has important effects on the vibronic spectra. The interesting point about these perturbations is that individual nitrogen
aggregates give little effect, but this is balanced by their relatively high concentrations, up to 4×10^{26} nitrogen atoms m^{-3}.

We consider two types of perturbation: First, the role of nitrogen in broadening optical transitions, and second, the quenching
of the luminescence of optical centers in diamond. Experiments on
these processes require gem quality, uncut, small diamonds if the
nitrogen effects are to be distinguished from spurious effects

caused by inclusions or by strain fields built into the specimens
when they are polished. Good data can be obtained using octahedral
type Ia diamonds of 1-mm thickness. These have the advantage that
they usually contain nitrogen in one dominant form, the A form
(Sec. III.B). Furthermore, the $\sim$1370-cm^{-1} peak, which indicates the
presence of platelets (Sec. III.E), is weak and the diamonds have
negligible dislocation densities [259].

Qualitatively, optical transitions in type IaA diamonds are
broadened as a statistical result of interactions between the A
aggregates and the optical centers. Since strains perturb optical
centers in diamond much more readily than do electric fields, we
expect the interaction of greatest importance to arise from the
strain field of the nitrogen. A given optical center will absorb
at a photon energy which is determined to a small extent by the strain
field it experiences as a result of all the nitrogen. A second optical center will, generally, be in a different state of strain and
so absorb at a different energy, and so on. An observed zero-phonon
line will be the envelope of all the zero-phonon lines occurring
at the individual optical centers.

The theory of strain broadening by point defects has been given
by Stoneham [260]. It can be applied with no adjustable parameters,
to optical centers in diamond if we treat the small aggregates of
nitrogen as point defects in an elastic continuum. Then the displacement field $\underline{u}$, and hence the strain tensor components, at a
distance $\underline{r}$ from an aggregate is given in terms of only one unknown,
M [261],

$$\underline{u} = \frac{M\underline{r}}{r^3} \qquad\qquad (41)$$

This unknown is related to the lattice expansion $\Delta V/V$ produced by a
fractional density ρ of aggregates:

$$\frac{\Delta V}{V} = 12 \; \pi \; M \; \rho \; \frac{1-r}{1+r} \qquad\qquad (42)$$

where r is Poisson's ratio, $0.1 < r < 0.3$. In practice we know

$\Delta V/V$ only for a given density of nitrogen, not density of aggregate [Eq. (9)], but it turns out that many properties of the broadening are given by $M\rho$ rather than ρ. If there are n nitrogen atoms per A aggregate then, from Eqs. (9) and (42),

$$M = (3.1 \pm 1.0)n \times 10^{-5} \text{ nm}^3$$

On the assumption of completely random distributions of strain sources and optical centers, Stoneham [260] has shown that a Lorentzian lineshape will be produced of full width at half height

$$\Gamma = \frac{1}{3} \rho \ |M| \ \pi \int_0^{2\pi} d\phi \int_0^\pi d\theta \ \sin \theta |\psi| \tag{43}$$

where the zero-phonon line is perturbed by strains e_{ij} according to the nondegenerate electronic state equation

$$\Delta h\nu = \sum_{ij} b_{ij} e_{ij} = \frac{M}{r^3} \ \psi(\theta,\phi) \tag{44}$$

Note that Stoneham's expression applies for random distributions in an elastic continuum. On average, $4\pi r^2 \rho \ dr$ nitrogen aggregates are assumed to be at a distance r from an optical center with $0 < r < \infty$.

Inspection of the theory shows that 99% of the linewidth is determined by those effects within a distance R of each optical center where [26] $R = 2 \ \rho^{-1/3}$. In type IaA diamonds, nitrogen concentrations are commonly 0.2 atomic % so that, for n = 2, a zero-phonon linewidth is determined by that part of the lattice within a distance R = 20 atomic spacings of each center. But a significant fraction of this volume is forbidden to the nitrogen aggregates in the sense that if nitrogen is found within a few atomic spacings of an optical center, a completely new center will be produced. The loss of this central region means that the highly perturbed optical centers present in the continuum theory are missing and so the observed zero-phonon line will have a lower intensity in the wings than a Lorentzian curve. Calculation [263] shows that a minimum radius of d $\sim$ 0.85 nm results in a bi-Lorentzian lineshape

$$I(\nu) = \frac{a}{\left[b + (\nu - \nu_0)^2\right]^2}$$

with a peak at ν_0 and width $\Gamma = 2\left[b[(2)^{\frac{1}{2}} - 1]\right]^{\frac{1}{2}}$. This lineshape is the most frequently observed shape in diamond, and it is pleasing to see that the strain broadening theory is not inconsistent with it.

The loss of the central region has little effect on the linewidth, and so Eq. (43) is still applicable. In particular, Γ is proportional to the aggregate concentration, as is usually observed experimentally, apart from a small residual component at zero-nitrogen concentrations (Fig. 38). The 2.463-eV (H3) line has a nitrogen-induced width at 0.1 atomic % nitrogen of

$$\Gamma_{\text{expt}} = (2.0 \pm 0.3) \text{ meV } [231, 263]$$

while the calculated value is

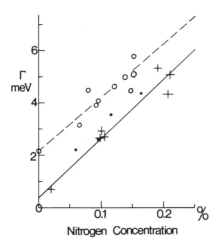

FIG. 38. Full width at half height of the 2.463-eV zero-phonon line at 80 K compared with mean nitrogen content in each specimen. Specimens used in the original work [263] fell into two categories shown by open circles and points. Data shown by points seem to have smaller residual broadening. Recent measurements using carefully chosen gem quality diamonds are shown by crosses. Note that the slopes of all three sets of data are the same, i.e., same nitrogen-induced broadening.

$$\Gamma_{calc} = (1.2 \pm 0.5) \text{ meV } [263]$$

The strain parameters b_{ij} have been derived from uniaxial stress
data by parameterizing it by a nondegenerate perturbation equation
[Eq. (44); see Sec. VI.C]. The 3.150-eV (ND1) linewidth due to
"random" distributions can be calculated from the nondegenerate
theory, if we use just one component of each symmetry of the strain
Hamiltonian [Eq. (32)]. At 0.2 atomic % nitrogen the widths are [231]

$$\Gamma_{expt} \sim 1.5 \text{ meV}; \quad \Gamma_{calc} = (0.7 \pm 0.4) \text{ meV} \qquad (46)$$

In both cases, the calculated values are only half the observed
values, probably because both these optical centers involve nitrogen
and so will be spatially correlated with it. On the other hand, the
nitrogen concentration used in the theory is a mean value through
the crystal: It is found by infrared absorption. Possibly no optical center in type IaA diamond is randomly distributed with respect
to the nitrogen. For example [231], some radiation damage centers
involve nitrogen directly (e.g., the 3.150-eV center), others have
intensities which increase with the nitrogen content (e.g., the
2.461-eV center), and still others show extrinsic structure caused
by migration to the nitrogen during the irradiation (e.g., the
1.673-eV center, Fig. 39). Usually any diamond with a high nitrogen
content ($\sim 10^{26}$ m^{-3}) will show strange lineshapes after irradiation
(Fig. 40). Heating to 900 K for a few seconds removes the highly
perturbed centers and gives nearly symmetrical shapes.

The peak frequency of a zero-phonon line is affected by the
nitrogen content. In all cases measured, the shift is to the red
with increasing nitrogen, although this need not be universally
true. Only the 2.463-eV lineshift has been studied in detail [263];
it is explicable by Stoneham's theory [260].

So far we have discussed absorption lines. Luminescence lines
show a nonlinear increase in width with nitrogen content, Fig. 41.
Also, the luminescence from diamonds of high nitrogen content is
less than one would expect from a diamond with the same concentration
of luminescent centers and a low nitrogen content [264]. Both these

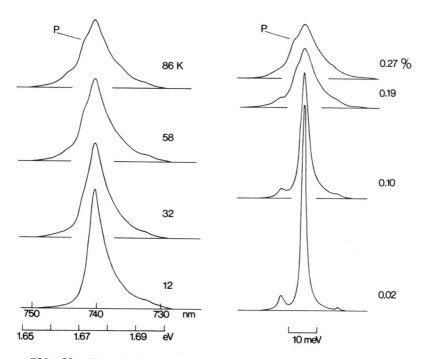

FIG. 39. Extrinsic structure of the 1.673-eV line caused by
high concentrations of nitrogen. At right, the 1.67-eV lines in a
series of uncut gem quality type IaA diamonds, measured at 80 K
after radiation damaging by 2-MeV electrons. With increasing nitrogen content, the lines broaden and show an extrinsic peak (P) to
low energy and a long tail to high energy. The independent 1.685-eV
line has been left on these spectra. All peaks have been superimposed for clarity. The extrinsic structure is shown more clearly
at left. As the temperature is lowered, P weakens, but most of the
high-energy tail remains. P primarily reflects a lifting of the E-
state degeneracy, while the high-energy tail primarily reflects the
splitting of the excited T-state of the 1.673-eV transition. The
degeneracy lifting is a result of nitrogen near to the 1.673-eV
centers [231].

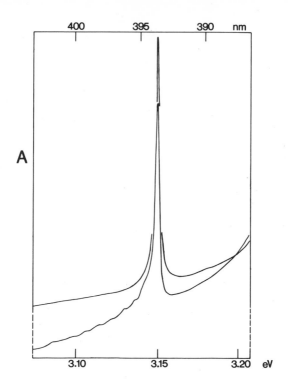

FIG. 40. The 3.150-eV (ND1) line after radiation damage is asymmetric, with a long tail to low energy, especially in diamonds with a high nitrogen content. The tail is structured in type Ib diamond (lower line). Annealing quickly makes the line more symmetric and stronger (upper line) In this case, the diamond was plunged into a furnace at 900 K for 30 sec and then quenched in cold water. Both spectra recorded to 80 K.

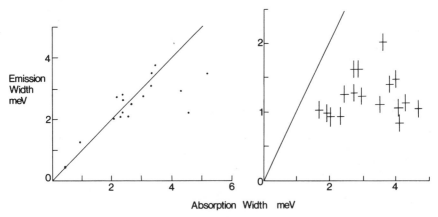

FIG. 41. The effect of luminescence quenching on the widths of zero-phonon lines is to make a luminescence line sharper than the absorption line in the same specimen. Data are for the 2.985-eV (N3) line at left and the 2.463-eV (H3) line at right, and were taken at 80 K. The points would lie on the lines if the widths were equal.

effects can be explained quantitatively if we assume that there can
be a transfer of energy from excited optical centers to the A aggregates of nitrogen where the energy is dissipated nonradiatively.
The only case treated in detail to date is the 2.463-eV (H3) center,
but other centers including the 2.985-eV (N3) center probably behave
in the same way [265].

First, we must establish the form of the coupling between the
optical centers and the A aggregates of nitrogen which leads to energy
transfer. The 2.463-eV center can make electric dipole allowed transitions; for example, its radiative lifetime is 15.5 ± 1 nsec [266,
267]. The A aggregates of nitrogen show no detectable absorption in
the visible spectrum (Sec. III.D). Any allowed absorption must be
by electric quadrupole or magnetic transitions. We postulate the
existence of an electric quadrupole moment; this is sufficient to
explain the phenomena under consideration.

Suppose we have N nitrogen aggregates at distances
r_i, i = 1,...,N, from a 2.463-eV center, r_i in units of the atomic
spacing. The coupling of the dipole moment of the center with the
quadrupole moment at each nitrogen atom enables energy to be transferred to the nitrogen. The probability that the 2.463-eV center
will decay radiatively is then reduced to

$$P = (1 + \frac{T_r}{T_n} \sum_{k=1}^{N} r_k^{-8})^{-1} \qquad (47)$$

since the probability of this nonradiative transfer process goes as
r^{-8} [268]. In Eq. (47), T_r is the radiative lifetime of the 2.463-eV
center and T_n is the lifetime of nonradiative transfer of energy from
the center to one nitrogen atom one atomic spacing away. (Obviously
T_n is a scaling factor without direct physical significance.)

For a given distribution of A aggregates, we can calculate the
probability of luminescence from Eq. (47) and the precise energy of
that luminescence from Eqs. (41) and (44). The calculation can be
repeated many times for different, randomly chosen configurations of
nitrogen. A histogram of the predicted luminescent zero-phonon line
can be built up. Many properties of the line can now be calculated.
We find that the width of the luminescence line will always increase
with increasing nitrogen concentration if the nitrogen is evenly

distributed in the diamond [26]. However, the luminescence efficiency decreases rapidly with nitrogen content [266] so that in practice the luminescence from a diamond of high nitrogen content is
dominated by any regions of low nitrogen it may have. The linewidth
then tends to saturate with increasing nitrogen, as in Fig. 41. The
ratio T_n/T_r appears from these considerations to be $T_n/T_r \sim 10^{-6}$
[262].

From the Monte Carlo calculations, one can also find the time
dependence of the luminescence decay. It turns out to be almost
exponential [266]:

$$I(t) \sim I(o) \exp \frac{-t}{T}$$

To overcome the problem of nitrogen inhomogeneity, we can express
the decay time T in terms of the luminescence linewidth Γ_ℓ. Somewhat suprisingly this complicated situation gives an almost linear
dependence of T on Γ_ℓ (Fig. 42). Figure 43 shows that this is observed experimentally. The radiative lifetime of the 2.463-eV center is about 15.5 ± 1 nsec [266, 267]. This is obtained from the

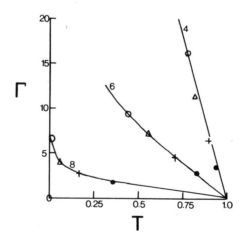

FIG. 42. Calculated decay times compared with the widths of
the emission lines for different values of the one parameter T_n/T_r.
Curves labeled 4, 6, 8 are for $T_n/T_r = 10^{-4}, 10^{-6}, 10^{-8}$. Repre-
sentative points are for fractional densities $\rho = 5 \times 10^{-4}$ (•),
10^{-3} (+), 2×10^{-3} (Δ), and 3×10^{-3} (o)

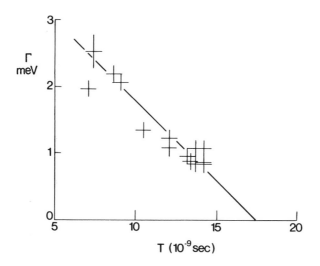

FIG. 43. The full width at half height of the 2.463-eV zero-phonon line measured in cathodoluminescence at 77 K compared with the decay time of the 2.463-eV luminescence in the same part of each diamond.

low nitrogen limit of the decay time on Fig. 43, remembering that there is a contribution to the linewidth of about 0.5 meV from the background effects (see Fig. 41). The shortest decay time actually measured, 7 nsec, was in a diamond of about $\rho = 2.7 \times 10^{26} \ m^{-3}$ A aggregates of nitrogen, assuming two nitrogen atoms per A aggregate. (Higher nitrogen concentrations quench the luminescence too strongly to measure T.) Using the ratio $T_n/T_r = 10^{-6}$ from the linewidth results, the decay time at $\rho = 2.7 \times 10^{26} \ m^{-3}$ is calculated to be 10 nsec in satisfactory agreement. Finally, the ratio $T_n/T_r \sim 10^{-6}$ is theoretically very reasonable [268].

We have seen that both the zero-phonon broadening and the luminescence quenching result from perturbations of the nitrogen on the optical centers. One novel consequence of this is that the zero-phonon line observed in absorption does not have the same peak energy as the same zero-phonon line observed in the luminescence spectrum of the same specimen. As the nitrogen concentration increases, the absorption line is progressively shifted, usually to the red. The most strongly perturbed centers tend to be those near nitrogen aggregates: Consequently, these have their luminescence strongly quenched. The emission zero-phonon line is produced by the remaining, less perturbed optical centers and so has a peak nearer the true, unperturbed energy of the center. In practice, the relative shifts are small but measurable; the 2.463-eV zero-phonon line has been seen at 0.3-meV higher energy in luminescence than in absorption in the same specimen [262].

In this section, we have seen that all the known perturbations of optical centers by the A aggregates of nitrogen can be understood quantitatively. The theory is based on the idea that the A aggregates contain very few nitrogen atoms each. If the aggregates were very large, for example related to platelets, there would necessarily be fewer of them. The energy transfer process, which is effective only over short ranges [note the r^{-8} dependence in Eq. (47)] would then be inoperative, and there would be no luminescence quenching. This is why the data in this section are very strong evidence in favor of small aggregates (Sec. III.E).

VIII. PHOTOCHROMIC EFFECTS

It has long been known that the absorption spectra of some diamonds can be changed by irradiating them with intense 365-nm Hg light [269, 270]. The effects are most pronounced in diamonds which have been subjected to radiation damage. For example, Dyer and du Preez [205] found that Hg light reduces the 3.150-eV (ND1) absorption and strengthens the 1.673-eV (GR1) band. Heating to 800 K reverses the process [205], or one can simply leave the diamond overnight in the dark. If other optical centers are present, they may also be modulated [206]. Recently, it has become clear that significant changes in the absorption spectra can occur with light intensities of the levels used in absorption spectroscopy [231]. It is essential to understand these processes if accurate absorption measurements are to be made.

From what has been said it is clear that the 1.673- and 3.150-eV bands are particularly sensitive. Now, to date there has been no observation of 3.150-eV luminescence. This led to the suggestion that the excited state of the center was unstable with respect to charge emission [204]. Farrar and Vermeulen [271] have detected these charges by measuring the photoconduction of the 3.150-eV band (Fig. 44). If we postulate that they can be captured by vacancies to produce optically active 1.673-eV centers, then the photochromic effect is understandable. This postulate is readily tested by measuring the growth of 1.673-eV absorption as the irradiating wavelength is scanned through the 3.150-eV absorption spectrum.

For these experiments, it is most convenient to use irradiated natural type Ib diamonds, since the equilibrium state of these specimens is to have a 3.150-eV absorption far stronger than the 1.673-eV absorption (Fig. 45a). For the results of Fig. 45b, the 1.673-eV zero-phonon line strength was measured at 80 K using very high spectral resolution. The incident light was filtered so that it had photon energies less than 2.1 eV. These have no effect on the 1.673-eV strength. A second beam of light was focused on the specimen through a low-resolution monochromator. When the energy of this light was between 3 eV and the upper practical limit of 3.4 eV, the 1.673-eV strength rose in an exponential decay [272], saturating, under these experimental conditions, in 3 hr. This 1.673-eV saturated value is plotted against the exciting wavelength in Fig. 45b

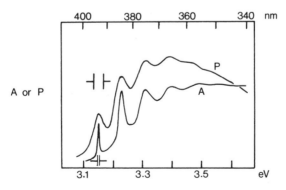

FIG. 44. Comparison of the photoconduction, P, and absorption, A, in a type Ia diamond containing 3.150-eV (ND1) centers. Data taken at 8 K by Farrar and Vermeulen [271].

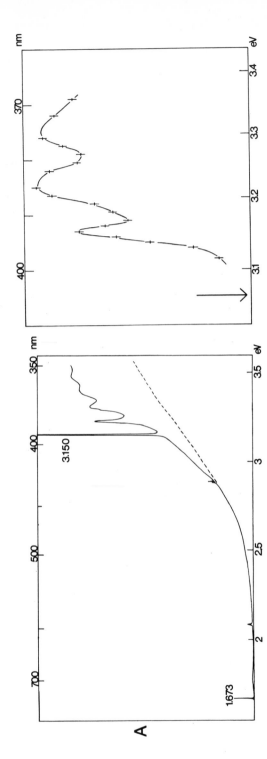

FIG. 45. (a) Absorption spectra of a type Ib diamond before (---) and after (——) radiation damage by 2-MeV electrons. Both spectra were taken at 80 K after prolonged exposure to ultraviolet light to enhance the 1.673-eV band. (b) Peak absorption in the 1.673-eV line after saturation by ultraviolet light plotted as a function of the exciting wavelength. The spectrum plots out the shape of the 3.150-eV band. No correction has been made for the absorption underlying the 3.150-eV band. The arrow shows the energy of light used to reverse the photochromic process and weaken the 1.673-eV band. It lies outside the 1.673-eV band but in the ultraviolet continuum of the 1.673-eV band.

and clearly maps out a low-resolution 3.150-eV spectrum confirming
that a simple charge transfer process is occurring.

Farrar and Vermeulen [271] have found that the ultraviolet absorption continuum associated with the 1.673-eV center is also photo-
conductive. Consequently, irradiation in these bands causes the
1.673-eV strength to decrease (Fig. 45b). We now see why, in a
type IIa diamond where the ultraviolet continuum is far stronger than
the 3.150-eV absorption, a general irradiation with ultraviolet light
can significantly increase the 3.150-eV strength; the favored transfer is from the many 1.673-eV centers to the few 3.150-eV centers.
The same irradiation in a type Ia diamond enhances the 1.673-eV
strength, since now the light is absorbed predominantly in the strong
3.150-eV band rather than the relatively weak ultraviolet continuum.

Absorption experiments in irradiated diamond should be made,
whenever possible, with the ultraviolet spectrum filtered out before
striking the diamond. Otherwise the changes described here must be
allowed for.

One interesting unknown in the photochromic process is the time
T, required for the 3.150-eV excited state to emit a hole. (The
emitted charge is evidently a hole in view of Farrar and Vermeulen's
identification of holes as being freed in the photoconduction of the
ultraviolet absorption, Sec. VI.B.) Since no 3.150-eV luminescence
is observed, T must be less than the radiative lifetime T_r of the
center. An experimental estimate is $T \leq 10^{-4} T_r$ [204]. Uniaxial
stress experiments have established the 3.150-eV transition to be
electric dipole allowed [204]. Therefore, $T_r \sim 10^{-8}$ sec and
$T \leq 10^{-12}$ sec. A simple calculation, in which all the electronic
energy is assumed to be transformed to kinetic energy of the outgoing
particle gave $T \sim 10^{-6} T_r$ [204]. This can be taken as a lower limit,
i.e., $10^{-14} \leq T \leq 10^{-12}$ sec. An excited state lifetime of this magnitude will produce appreciable lifetime broadening of the zero-
phonon line; for example, $\hbar/10^{-12}$ sec = 0.7 meV. This may be the
reason why the sharpest 3.150-eV line observed so far is 2 meV in
contrast to 0.5 meV for the 2.463-eV line [231], especially since
the 3.150-eV line is not expected to be very sensitive to
inhomogeneous broadening [Eq. (46)].

IX. DONOR–ACCEPTOR PAIR SPECTRA

In the previous two sections, we have discussed optical properties
involving more than one optical center. Another two-center property
is the luminescence emitted by the recombination of an electron on
a donor center with a hole on an accepter center. These donor-
acceptor pair spectra are well established in several materials,
notably GaP [273]. The role they play in diamond is not yet fully
understood. Since a detailed study of the spectra is currently in
progress in this laboratory, we only sketch their properties here
with a minimum of speculation.

If one neutral donor is at a large distance r from one neutral
acceptor, the energy emitted by the recombination of the electron
and hole is

$$h\nu = E_g - E_a - E_d + \frac{e^2}{4\pi\varepsilon r} - \frac{a}{r^6} \qquad (48)$$

The term a/r^6 represents the polarization interaction of the neutral
centers, a Van-der-Waals type attraction. For large r it has a
smaller magnitude than $e^2/4\pi\varepsilon r$. The criterion for large distances
is that r must greatly exceed the Bohr radius of the least well-bound
particle so that there are negligible complications from the overlap
of, say, the hole wavefunction with the donor site [274]. In the
best situations, long series of zero-phonon lines are seen corresponding to different separations r [275]. However, nonrandom distributions of the centers and electron-lattice interactions at the centers will destroy the discrete line nature of the spectra, producing
broad bands. To recognize the bands as pair spectra, we must then
rely on such properties as the spread of luminescence decay times
across the band produced by the rapid reduction in the speed of recombination as r increases [276], or the change in spectral shape on
changing the power of the excitation used to generate the spectrum
[277].

Measurements of this type led Dean [111] to suggest that the
blue cathodoluminescence emitted from natural diamond is a result of

pair spectra. So far in this chapter we have only met one naturally
occurring acceptor center in diamond, the effective mass acceptor
of E_a = 0.368 eV (Sec. III.G). The acceptor involved in pair spectra
is presumably this one. Some evidence in favor of this comes from
the thermally activated quenching of the luminescence reported by
Dean and Male [278]; the activation energy was 0.36 ± 0.02 eV. To
identify the donor, we note that the low-energy cut-off of the luminescence is just above 1 eV, in as much as it can be defined.
According to Eq. (48), the lowest energy electronic transition occurs
as $r \to \infty$ at $\xi_m = E_g - E_a - E_d$. If this transition forms a vibronic
band, ξ_m will lie at its centroid, and the low-energy cut-off will
be a few hundred millielectronvolts to lower energy. Allowing for
these uncertainties it seems likely that $E_d \sim$ 3.5 to 4.0 eV. Certainly E_d > 2 eV. This suggests that the donor is the A aggregate
of nitrogen (Secs. III.D and III.F). In type IaA diamond, we then
expect the donor-acceptor pair spectra to be difficult to understand
because of the high density of donors; things should be simpler in
type IIb diamonds.

     In natural type IIb diamonds, the blue cathodoluminescence takes
the form of a broad band with a high-energy cut-off near 3.3 eV at
80 K and a peak near 2.8 eV. Close inspection of the band using
time-resolved spectroscopy shows that there is more than one component present, with luminescence decay times of 5 to 8 nsec [232].
At lower energy, a weak green cathodoluminescence can just be detected. In synthetic type IIb diamonds, this green luminescence is
strong, dominating the blue luminescence, as in Fig. 10 of [111].
The green peak shows the characteristic properties of donor-acceptor
pair spectra [279]. For example, the peak of the luminescence shifts
from 2.2 to 2.1 eV as the time delay between excitation and observation is increased from 0 to 10 msec [232]. Although the green peak
is far stronger than the blue peak in synthetic diamond, the blue
peak can still be observed by gating the photomultiplier detector
so as to detect only the luminescence emitted within a few nanosec-
onds of the excitation pulse. The blue peak then appears to be

similar to that in natural type IIb diamond [232]. The differences
between the green and blue peaks seem to be consistent with Dean's
idea [111] that the donors and acceptors tend to be at very close
separations in natural type IIb diamond, but that in synthetic
diamond they are frozen in more nearly random configurations as a
result of the relatively short synthesis times [113].

The similar donor-acceptor pair spectra observed in natural and
synthetic type IIb diamond means that the same donors and acceptors
are active in both groups of diamonds. We have suggested that the
donor is the A aggregate of nitrogen and so the implication is that
A aggregates are found in synthetic diamond as well as the single
substitutional form. Much stronger evidence that these aggregates
can occur in synthetic diamond comes from the frequent observation
of the 2.463-eV (H3) band in their cathodoluminescence spectra, for
the 2.463-eV center is undoubtedly a radiation damage center trapped
at an A aggregate of nitrogen (Sec. VI.C). Klyuev et al. [90] have
also reported A aggregates in synthetic diamond. It may seem strange
that a minor form of nitrogen, A aggregates in synthetic diamond,
can apparently dominate the luminescence. But it must be remembered
that the intensity of luminescence is not related in any simple way
to the concentration of optical centers producing the luminescence
(e.g., Sec. VII). There are well-established cases in diamond where
the luminescence is caused by concentrations of centers that are in-
significantly small by any other standards [280]. This is one of
the main difficulties in interpreting luminescence spectra of diamond.

One interesting feature of donor-acceptor pair spectra is that
there is no obvious luminescence from pairs involving the single
nitrogen atoms characteristic of type Ib diamond. This would lie
in the ultraviolet, according to Eq. (48) with E_d = 1.7 eV (Sec. III.F).

There is still a great deal to learn about donor-acceptor pair
spectra in diamond, not only for their own sake but also to understand
the role they play, if any, in the thermoluminescence spectra of
diamond [281, 282]. Hopefully some of the remaining problems will
soon be clarified by measurements currently in progress in this

laboratory, and for this reason, we do not discuss these important transitions any more here.

X. FINAL REMARKS

We have seen that the optical properties of diamond are now known, to a large extent, for photon energies covering three orders of magnitude, from 30 meV to 31 eV. But this knowledge is almost entirely empirical; as yet, there is very little understanding of the optical properties from first principles. This is particularly disturbing since the atoms involved, mainly carbon and nitrogen, are of such importance in many other fields. In fact, the main motivation for basic diamond research from a purely scientific point of view must be that diamond forms a simple, large molecule whose constituents it should be possible to define precisely. Unfortunately, there are reliable models for only two centers: The single nitrogen atom of type Ib diamond (Sec. III.F) and the acceptor center in type IIb diamond (Sec. III.G through III.I). Other models are partially established; for example, the 1.673-eV (GR1) center is a vacancy in some unknown charge state (Sec. VI.A), the 3.150-eV (ND1) center is probably a tetrahedral interstitial nitrogen atom (Sec. V.B), and the 2.985-eV (N3) center is possibly three substitutional nitrogen atoms in a C_{3v} configuration (Sec. V.C). But the most abundant impurity, the A aggregate of nitrogen is of unknown structure, although we may guess it contains two nitrogen atoms (Sec. III.E). Nor is the B nitrogen aggregate understood. It seems to be a near relative of the A aggregate in that they produce the very similar 2.463- and 2.499-eV (H3 and H4) bands when they trap the same species of radiation damage center (Sec. VI.C). Yet the B nitrogen has no strong ultraviolet absorption in contrast to the A nitrogen (Fig. 9). The structures of the A and B nitrogen aggregates present perhaps the most important experimental problems at the moment, since they are the key to understanding the optical properties of type Ia diamond.

Many other points of ignorance concerning the optical properties have been mentioned in the relevant sections and need not be repeated here. But two whole fields of study have not been mentioned yet: Electron paramagnetic resonance and thermoluminescence. In recent years, the epr of centers in diamond have received less attention than have their optical transitions. Recent papers are given in Refs. 221 and 283. It is still not clear how most of the epr centers relate to the optical centers [284]. A major study of this is required.

Thermoluminescence has been ignored in this chapter on the grounds that it is an oversophisticated tool for probing a material like diamond for which one has little real knowledge of the impurities. For example, it is well established that the thermoluminescence glow curves of type IIb diamonds contain peaks at 150 and 250 K with the luminescence being emitted mainly near 1.8 eV [281, 282, 285]. The same basic mechanism must be at work in all type IIb specimens. Now we know from electrical studies that type IIb diamonds contain compensated donors and that these may play a role in the thermoluminescence [281]. As yet there is no independent way of identifying the donors, partly because of their low concentration, $\sim 10^{21}$ m^{-3}, and partly because they are normally charge compensated and so will have different properties from, e.g., the A aggregates of nitrogen even if they have the same molecular structure. The thermoluminescence may, therefore, be caused by structures that cannot be independently assessed, and we are free to speculate about it with few external constraints.

A similar, but better established, situation arises in the cathodoluminescence spectra of type II diamonds which have been subjected to radiation damage and annealing. The visible luminescence is dominated by the 2.156- and 3.188-eV bands. When the absorption spectra of these diamonds are examined, there is hardly any trace of these bands. Since they are both most probably electric dipole transitions, their optical centers must be present in concentrations below 10^{-v} m $^{-}$. Again the luminescence is being determined by centers which are present in trivial concentrations. Generally, it is far more difficult to understand luminescence spectra from type II than type I diamonds, for in type I specimens there is the advantage of being able to make (absolute) absorption measurements of the dominant centers. This underlies the theory of luminescence quenching in Sec. VII.

We have not yet discussed any applications of the optical properties of diamond. These are disappointingly few. There is the important, yet trivial, application of diamond in jewelry, exploiting its high refractive index [35]. The same phenomenon was used briefly in the nineteenth century in making high-power, single-lens magnifying glasses from diamond [286]. These magnifiers had very low chromatic aberrations as a result of the relatively low dispersion of diamond. Visible photoluminescence of diamond from the 2.985-eV (N3) band is used to sort diamonds into, hopefully, the relatively pure type II diamonds and others, so that diamonds of the highest thermal conductivity can be selected. For reasons such as given in Sec. VII, the technique has real limitations.

Otherwise optical studies of diamond remain a matter for pure science. There is now a remarkably long history of these studies. For example, triboluminescence of diamond was observed as long ago as 1662, by Boyle [287]. Thermoluminescence studies can be traced back to the twelfth century [288]; again Boyle was one of the more scientific investigators (in 1672 [287]). Newton gave the refractive index as 2.439 in 1704 [289]. More recently, Crookes described the cathodoluminescence of diamond in 1879 [290] and later discovered the 2.463-eV (H3) band, while Walter observed the 2.985-eV (N3) absorption band in 1891 [215]. Radiation damage studies date from the beginning of this century [291]. It was probably the aesthetic attraction of diamond that led to its being used in these pioneering studies. The present level of activity shows that it has not lost its appeal.

ACKNOWLEDGMENTS

I thank A. T. Collins and E. C. Lightowlers for many enlightening
discussions; R. J. Caveney, W. F. Cotty and F. A. Raal for providing
the specimens used for the original parts of this chapter; and
de Beers Industrial Diamond Division for their continuing support
of my work. I am also grateful to P. Silverthorne, P. Walker, and
my wife for preparing the typescript. Figures 13 and 14 are reproduced with the permission of S. D. Smith and the Institute of
Physics, and Figs. 15, 16, and 17 with the permission of E. C.
Lightowlers and *The Physical Review*.

APPENDIX: ABSORPTION, LUMINESCENCE, AND PHOTOCONDUCTION TRANSITIONS IN DIAMONDS

Table A1 lists the broad bands seen in diamonds. Omitted from the
table are the defect-induced one-phonon transitions used to define
the types of diamond (see Figs. 3, 10, and 13). Also omitted are the
two- and three-phonon intrinsic transitions (see Fig. 1). Table A2
lists optical transitions with zero-phonon lines or well-defined
thresholds. Omitted from this table are many absorption and cathodoluminescence lines seen near 2.4 eV in type Ia diamonds after radiation damage and various annealing treatments, since their generality
is not yet established.

The main group of transitions falling in neither table are the
absorption lines which are caused by photon absorption with the
creation of one phonon. For example, the 1450-cm^{-1} (180 meV, H1)
line of typical full width at half height of 4 cm^{-1} at 300 K, which
is seen after radiation damage and annealing, or the 1405- and
3107-cm^{-1} (174, 385 meV) lines of typical widths $\sim$4 cm^{-1} at 300 K
seen in type Ia diamonds in their natural state. Angress et al.
[297] list other radiation damage-localized mode lines and Sutherland
et al. [25] give lines at 1520 and 1540 cm^{-1} (189 and 191 meV)
in natural type Ia diamonds.

TABLE A1

Broad Bands in Diamond[1]

eV (nm)	Description
1.2 (1030)	C. IaA. Band of half-width ∿330 meV at 120 to 310 K. Intensity correlates with strength of the ∿1370-cm⁻¹ line (Sec. III.C) [235]. See Fig. 1 of [235].
1.2 (1030)	C. Ia, IIa, IIb. Appearance like that of a vibronic band with Huang-Rhys factor S ∿ 9 at 50 K and with well-defined phonon energy ℏω = 53 meV, intrinsic phonon linewidth ≲ 1 meV. See Fig. 12 of [235].
1.8 (690)	C. No type dependence. Band of full-width at half-height 350 meV at 50 K, with a lot of fairly weak structure on it. See Fig. 11 of [235].
2.1 (590)	C. Synthetic diamond. Band of half-width 450 meV at 80 K, could vary. Donor-acceptor spectrum. See Sec. IX.
2.23 (555)	A. Diamond type unknown (or irrelevant?) in view of the rare occurrence of band. Band half-width 350 meV, insensitive to temperature. Raal [19] suggests band could be Mn dependent. See Fig. 1 of [19].
2.25 (550)	A. Rare band of half-width 200 meV at T < 300 K. See Fig. 4.9 of [207].
2.6 (447)	A. Some type I diamonds. Half-width 350 meV at 300 K, some structure on low energy side at 80 K. See Fig. 4.4 of [207].
2.8 to 3.1 (443 to 400)	C. Ia, IIa, IIb. Band of half-width ∿450 to 800 meV at 80 K, but could vary. Donor-acceptor pair with fairly close separation? Radiative lifetime 5 to 8 nsec [232]. See Sec. IX.

[1]The first column gives the energy (wavelength) of the approximate peak of the band. The second column lists the means of observation (A, absorption; C, cathodoluminescence), the types of diamond in which the band can be seen and any general comments.

TABLE A2

Broad Bands Listed by Zero-Phonon Energy and Fundamental
Edge Transitions of Diamond

The first column gives the energy (wavelength) of the transition at 80 K in a diamond with small perturbations. The second column gives the name of the band. In the third column we list:

1. Means of observation: A, absorption; C, cathodoluminescence; L, photoluminescence; L', luminescence excitation; P, photoconductivity.

2. Types of diamond in which the transition can be seen.

3. Means of production.

4. Probable models for the center.

5. Dominant phonon energies, hω, and Huang-Rhys factor, S, defined from intensities of zero-phonon line and total band at 80 K.

6. Zero-phonon lines that occur at the same optical center.

7. Reference to diagram of spectrum. Unfortunately, several interesting bands are only described in Ph.D. theses.

eV (nm)	Name	Description
0.305, 0.342, 0.347, 0.350, 0.363 (4065, 3625, 3573, 3542, 3415)		A, P (except 0.305 eV). IIb. Internal transitions at acceptor center, probably boron: Sec. III.I. Photoconduction arises from photothermal ionization: Sec. III.G. hω ∼ 165 meV. Many other lines observed in spectrum, see Charette [4], Smith and Taylor [118], Crowther et al. [131]. Related to 5.356- and 5.368-eV lines. See Figs. 13 through 17, Secs. III.G and III.I.
0.77 (1610)		A. Naturally occurring. hω = 60 and 80 meV. Very rarely seen. See Fig. 4.10 of [207].

1.25 (992)	H2	A. Ia. Radiation damage and annealing at 800 to 900 K. Higher temperature anneal does not produce band, although once formed it is stable at 1200 K. Presumably a radiation damage center trapped at an A aggregate. $\hbar\omega$ = 70 meV, $S \sim 4$. See Fig. 9.6 of [207].
1.264 (980.9)		C. All types of diamond? $\hbar\omega \sim 60$ meV, $S \sim 1$. See Fig. 50 of [292].
1.403, 1.440 (883.7, 885.6)	1.4 eV	C. Ib. Naturally occurring. $\hbar\omega$ = 60, 165 meV, $S \sim 1.6$. Doublet splitting occurs in ground state, doublet components have same strength. Internal transition at single substitutional nitrogen atom? [235]. Radiative lifetime 20 nsec [232]. See Figs. 6 through 8 of [235].
1.50 (825)	N1	A. Ia. Naturally occuring. Weak absorption system of small temperature dependence, somewhat similar in appearance to 2.596-eV (N2) line (Fig. Al). See Fig. 2 of [48].
1.665, 1.673 (744.6, 741.1)	GR1	A, C, L, L'. All diamonds. Radiation damage product destroyed by annealing at 900 K in type Ia diamond or up to 1200 K in type II. Lines are, respectively, A to T and E to T transitions at a T_d center [233, 234], most likely the vacancy. The T state is common to both lines: Fig. 29 $\hbar\omega$ = 40?, 90, 165 meV, $S \sim 3.5$: Fig. 27. Lines usually weak in irradiated natural type Ib diamond. The 1.673-eV center is very important in photochromic effects: Sec. VIII. Radiative lifetime 3 nsec [232]. Related to 2.881-, 2.888-, 2.902-, 2.904-, 2.958-, 2.960-, 2.976- and 2.982-eV lines and 2.997-eV multiplet See also Fig. 39 and Sec. VI.A.
1.685 (735.8)		A. All diamonds. Radiation damage product, destroyed by heating to about 700 K. Related to 1.860-eV line. Observed just to higher energy than 1.673-eV line, freezes out on lowering temperature at same rate as 1.665-eV line [231], but is independent of these lines. See Fig. 39.

TABLE A2 (Continued)

eV (nm)	Name	Description
1.762 (703.6)		C. Ib. $h\omega = 33$ meV. See Fig. 48 of [29].
1.860 (666.6)		A. All diamonds. Radiation damage produce, destroyed by heating at 700 K. Related to 1.683-eV line. See Fig. 27.
1.883 (658.4)		A. Ib synthetic diamond. Radiation damage product. Related to 1.906-, 1.914-, and 1.943-eV lines. See Fig. A1.
1.906, 1.914 (650.5, 647.8)		A. IB synthetic diamond. Radiation damage product. Related to 1.883- and 1.943-eV lines. See Fig. A1.
1.908, 1.924 (649.8, 644.4)		A. IA. Radiation and annealing, conditions unknown. Doublet splitting in excited state. Related to 2.092-, 2.321-, 2.341-, 2.712-, and 2.718-eV lines. See Fig. 2 of [255].
1.943 (638.1)		A. Ib synthetic diamond. Radiation damage product. Related to 1.883-, 1.906-, and 1.914-eV lines. See Fig. A1.
1.945 (637.4)		A. L, L'. Radiation damage and annealing at 900 K. A to E transition, trigonal center; nitrogen vacancy pair? See Sec. VI.B. $h\omega = 65$ meV, $S = 3.65$. Very readily produced; gives diamonds a characteristic mauve color. Not been seen in cathodoluminescence. See Figs. 31 and 32.
1.979 (626.4)		A. Ib. Radiation damage. Anneals out at 800 K.
2.086 (594.3)		A. Ia, Ib. Radiation damage and annealing. $h\omega = 85$ and weak 165 meV, $S = 2.1$. Very rapid shifts under stress [203]. See Figs. 21 through 24 and A2.
2.092 (592.6)		A. Ia. Radiation damage and annealing, conditions unknown. Related to 1.908-, 1.924-, 2.321-, 2.341-, 2.712-, and 2.718-eV lines. See Fig. 2 of [255].

2.156 (575.0)		A, C, L'. All diamonds. Radiation damage product. Band strength is increased substantially by annealing, e.g., at 1000 K for 2 hr. Band strength also increases, irreversibly, as a result of the cathodoluminescence beam at 80 K, an example of athermal annealing $h\omega$ = 55 meV, S ∿ 3.3. Radiative lifetime ∿29 nsec [232]. See Fig. 3 of [293].
2.305 (537.9)		C. Ia. A sharp line superimposed on the 2.463-eV (H3) cathodoluminescence band, apparently an internal transition at the same center. See description of 2.463-eV line.
2.321, 2.341 (534.2, 529.6)		A. Ia. Radiation damage and annealing, conditions unknown. Doublet splitting in excited state. Related to 1.908-, 1.924-, 2.092-, 2.712-, and 2.718-eV lines. See Fig. 2 of [255].
2.365 (524.])		A. Ib. Radiation damage product. Light near 2.365 eV bleaches the center: Warming to room temperature restores it [294]. Anneals out at 900 K.
2.400 (516.6)	M1	A. All diamonds. Radiation damage product, stable to at least 1100 K. $h\omega$ ∿ 21 meV? S << 1. See Fig. A3.
2.417 (512.9)		A. Ia. Radiation damage and annealing at 900 K. Monoclinic I center. Sec. VI.C. See Figs. 35 and 36.
2.445 (507.1)	M2	Details as for 2.400-eV line.
2.462 (503.6)	3H	A, L, L'. All diamonds. Radiation damage product, especially strong after high energy electron radiation (>>2 MeV). In type IaA diamonds strength grows with increasing nitrogen content and line is symmetric with tail to lower energy. Very easily produced in type Ib diamonds [231]. $h\omega$ = 66 meV, S ∿ 0.8. Anneals out at 700 K in diamonds of high nitrogen content, stable to higher temperature in type IIb diamond. Rhombic I symmetry [296]. See Figs. A3 and A4.

TABLE A2 (Continued)

eV (nm)	Name	Description
2.463 (503.4)	S1	L. Diamond type unknown. $h\omega$ = 45 and 69 meV. According to [295], a weak oscillatory series of phonon sidebands up the side of a continuum of luminescence.
2.463 (503.4)	H3	A, C, L, L'. IaA. Radiation damage and annealing at 900 K. Radiation damage center (3.150 eV?) trapped at A aggregate of nitrogen: Sec. VI.C. $h\omega$ = 41 and 152 meV, $S \sim 3$. Rhombic I center. Related to 2.305- and 3.368-eV lines. Radiative lifetime 15.5 nsec. See Figs. 35 through 38.
2.499 (496.1)	H4	A, C, L, L'. IaB. Radiation damage and annealing at 900 K. Radiation damage center (3.15 ev?) trapped at B aggregate of nitrogen: Sec. VI.C. $h\omega \sim 40$, 154 meV, $S \sim 3$. Monoclinic I center. Similar properties at 2.463-eV line. See Figs. 35 and 36.
2.500 (495.9)		A. IIa. Radiation damage product, stable to annealing at 1100 K. Doublet of small intensity. See Fig. A3.
2.520 (492.0)		C. IIb. Naturally occurring weak line commonly observed in type IIb diamonds. See Fig. 10 of [235].
2.523, 2.536 (491.4, 488.9)		C. Ib. Radiation damage product. Related to 2.560- and 2.586-eV lines. Gives bright green luminescence. See Fig. A5.
2.533 (489.5)		A. Ib. Radiation damage product. $h\omega$ = 37 and 74 meV. $S \sim 1$.
2.543 (487.5)	TH5	A. IIa. Radiation damage and annealing at 900 K. Clark et al. proposed center was a pair of 1.673-eV centers [152]. No published diagram.
2.560, 2.586 (484.3, 479.4)		C. Ib. Radiation damage product. Related to 2.523-, 2.536- and 2.586-eV lines. The 2.560- and 2.586-eV lines thermalize according to I(2.560)/I(2.586) = (0.61 ± 0.01) exp(-E/kT) where E = 26 meV. Gives bright green luminescence. See Fig. A5.

Energy eV (nm)	Label	Description
2.596 (477.6)	N2	A. Ia. Naturally occurring. Related to 2.985- and 3.603-eV lines [48]. Structure always broad, cooling to 4 K does not sharpen lines appreciably relative to 80 K. See Fig. A6.
2.638 (470.0)	TR12	A, L, L'. IIa (also seen weakly in Ia: Fig. 36). Radiation damage destroyed by annealing at 900 K. $\hbar\omega \sim 33$ meV, $S \sim 0.6$, in absorption. One-phonon sideband sharp, intrinsic full-width at half-height 3 meV, and called TR13 in [48]. In absorption, zero-phonon intensity falls off in accordance with Eq. (29) and $\hbar\omega = 33$ meV, $S = 0.6$. But luminescence shows $\hbar\omega \sim 70$ meV. Center suggested to have monoclinic I symmetry but evidence not decisive [296].
2.712, 2.718 (457.2, 456.1)		A. Ia. Details as for 1.908- and 1.924-eV lines. See Fig. 2 of [255].
2.725 (455.0)		C. IIb. Weak, naturally occurring cathodoluminescence line commonly observed in type IIb diamond. See Fig. 10 of [235].
2.777 (446.5)	TR14	A. IIa. Radiation damage product. See Fig. 7 of [48].
2.788 (444.7)	TR15	A. IIa. Radiation damage product. See Fig. 7 of [48].
2.817 (440.1)	TR16	A. IIa. Radiation damage product. See Fig. 7 of [48].
2.831 (437.9)	TR17	A. IIa. Radiation damage product. See Fig. 7 of [48].
2.833 (437.6)		C. IIb. Weak, naturally occurring line commonly observed in type IIb diamonds.
2.881, 2.888 (430.3, 429.3)	GR2,3	A, L', P. All diamonds. Radiation damage product. Transitions are from the E ground state of the 1.674-eV line to separate T excited states: Sec. VI.A and Fig. 29. Related to 1.665-, 1.673-, 2.902-, 2.940-, 2.958-, 2.960-, 2.976-, 2.982- and 2.997-eV multiplet. See also Figs. 27 and 28 and Sec. VIII.
2.902 (427.2)	GR4	A. Another internal transiton at 1.673-eV center. See Figs. 28 and 29.

TABLE A2 (Continued)

eV (nm)	Name	Description
2.918 (424.9)		A. Ib. Radiation damage and annealing, e.g., at 900 K for 2 hr. $\hbar\omega \sim 55$ meV. See Fig. 31.
2.920 (424.6)		A. IaB. Radiation damage and annealing at 900 K for 2 hr. $\hbar\omega = 12$ meV. Related to 2.499 eV (H4)? See Fig. A7.
2.940 (421.7)	GR5	A, P. Another internal transition of 1.673-eV center. See Figs. 28 and 29.
2.958, 2.960 (419.1, 418.9)	GR6	A, L', P. Further internal transitions at 1.673-eV center. See Figs. 28 and 29.
2.976, 2.982 (416.6, 415.8)	GR7	A, P. Further internal transitions at 1.673-eV center. See Figs. 28 and 29.
2.985 (415.3)	N3	A, L, L'. Ia diamonds, especially those with strong B absorption bands (but no quantitative correlation). Naturally occurring center. A_1 to E transition at C_{3v} center, possibly three nitrogen atoms bonded to a common carbon atom. $\hbar\omega = 93$, 165 meV, $S = 3.45$. Related to 2.596- and 3.603-eV lines. Radiative lifetime 40 nsec [267]. See Figs. 19, 22 through 26, and 41 and Sec. V.C.
2.997 to 3.007 (413.7 412.3)	GR8	A, P. Further internal transitions of 1.673-eV center. A multiplet of unresolved structure. See Figs. 28 and 29.
3.04 (407.8)	R9	A, P. Ia, IIa. Radiation damage center. See Fig. 2 of [247].
3.150 (393.6)	ND1, R10	A, P. Ia, Ib, IIa. Radiation damage product resulting from damage at room temperature. No absorption produced in Ia diamond held at T < 210 K during radiation [207]. A to T transition at T_d center, probably interstitial nitrogen atom [204]. $\hbar\omega = 80$ meV, $S = 3.18$, very simple oscillatory vibronic band: Fig. 20. Excited state unstable to charge emission; no luminescence yet

eV (nm)	Label	Description
3.188 (388.9)		A, C. All specimens. Radiation damage and annealing at 1000 K for 2 hr. $\hbar\omega = 80$ and 165 meV $S = 1.82$: Fig. 21. Much structure often (always?) superimposed near optical phonon sideband: Fig. A8. Very readily seen in cathodoluminescence but weak in absorption (see Fig. 31). See also Figs. 22, 24, and 26 and Sec. V.A.
3.368 (368.2)		A, L'. Ia.A. Radiation damage and annealing at 900 K. Transition at 2.463-eV (H3) center probably from the same ground state as the 2.463-eV line. See Fig. 4 of [219].
3.603 (344.1)	N4	A, L'. Ia. Naturally occurring center. Transition at 2.985-eV (N3) center from same A_1 ground state to an excited A_1 state [219]. See Fig. 7 (type IaB spectrum).
3.765, 3.933 (329.3, 315.2)	N5,6	A. IaA. Naturally occurring centers. Possibly electronic transitions at the A nitrogen aggregate. Also related to 4.050- and 4.19-eV lines. See Fig. 8 and Sec. III.D.
3.988 (310.9)	R11	A. IIa. Radiation damage center. See Fig. 6 of [48].
4.050, 4.19 (306.1, 296)	N7,8	A. IaA. Naturally occurring centers. Possibly electronic transitions at the A nitrogen aggregate. Related also to 3.765- and 3.933-eV lines. See Fig. 8 and Sec. III.D.
4.137, 4.20, 4.27, 4.316, 4.355, 4.562 (299.7, 295, 290, 287.2, 284.7, 271.8)		A. Ib. Naturally occurring transitions at isolated substitutional nitrogen atoms. See Figs. 11 and 12.
4.437 (279.5)		C. IIa, IIb. Radiation damage center. $\hbar\omega \sim 237$ meV as for 4.582-eV line, but not related to it. See Fig. A9.

TABLE A2.(Continued)

eV (nm)	Name	Description
4.582 (270.6)	5RL	C. IIa, IIb. Radiation damage center. $\hbar\omega$ = 58 and 237 meV, localized mode. For this mode, S ∼ 1.5. Possibly lines at 4.388 and 4.582 eV are weak-phonon sidebands of localized modes $\hbar\omega$ = 194 and 175 meV. 5RL is one of the very few vibronic bands in diamond showing localized modes of vibration. (See also 4.698- and 4.777-eV lines.) These localized mode sidebands are essentially of the widths expected for zero-phonon lines and so could be confusing if a full spectrum is not available. See Fig. A9.
4.676, 4.639 (265.1, (267.3)		C. IIa, IIb. Two weak, unrelated lines seen in specimens which have probably suffered radiation damage as in Fig. A9.
4.698 (263.9)	2BD(G)	C. IIb. Possibly radiation damage product.$\hbar\omega$ = 220 ± 1 meV with S ∼ 2, localized mode. See Fig. 43 of [292].
4.777, 4.781, 4.803, 4.830 (259.5, 259.3, 258.1, 256.7)	2BD(F)	C. IIb. Possibly radiation damage product. $\hbar\omega$ = 210 ± 1 meV with S ∼ 1, localized mode. See Fig. 43 of [292].
4.846, 4.855 (255.8, 255.4)	T,T'	C. IaB. Naturally occurring center showing a zero-phonon doublet, T, T' with T' about half intensity of T at 100 K (excited state splitting?). Vibronic spectrum formed with well-defined phonons of $\hbar\omega$ = 148 meV, S ∼ 1. See Fig. 29 of [292].
4.986 (248.7)	N	C. IaB. Naturally occurring center with S << 1. See Fig. 29 of [292].
4.999 (248.0)	M	C. IaB. Naturally occurring center with S << 1. Not related to nearby similar 4.986-eV line. See Fig. 29 of [292].
5.032, 5.042 (246.4, (245.9)	L,L'	C. IaB. Naturally occurring pair of lines, L' ∼ 10% intensity of L at 100 K. Possibly $\hbar\omega$ = 141 meV, S ∼ 1 (giving peak S on Fig. 29 of [292] as one-phonon sideband). See Fig. 29 of [292].

Energy (eV)	Label	Description
5.092 (243.5)	K	C. IaB. Naturally occurring center. (See Fig. 29 of [292]).
5.135 (241.4)	E_0	C. IIa, IIb. Naturally occurring or radiation damage center. $\hbar\omega$ = 165 meV, S ~ 0.5. Low-energy threshold of line is typically at 5.135 eV, peak near 5.18 eV at 80 K. See Fig. 8 of [116] and Sec. III.H.
5.251, 5.261 (236.1, 235.7)	N9	A, C, L, L', P. IaB. Naturally occurring center whose strength is proportional to strength of B infrared bands but much weaker than expected for an electric dipole transition at the B nitrogen [47]. $\hbar\omega$ = 76, 141, and 163 meV and 141 meV dominant. S = 2.16. Doublet splitting is in excited state. See Figs. 21 and 22 and Secs. III.D and V.A.
5.258 (235.8)	S_0	C. IIb synthetic diamonds. Naturally occurring center. See Fig. 9 of [116] and Sec. III.H.
5.246, 5.253 (236.3, 236.0) 5.268, 5.275 (235.3, 232.7) 5.322 5.329 (233.0, 232.7)	A, A' B, B' C, C'	C. IIa, IIb. Intrinsic features of diamond. Produced by recombination of a free electron at a conduction band minimum with free hole at valence band maximum with emission of a wavevector conserving phonon. Energies quoted are for thresholds of peaks. See Fig. 18 and Sec. IV.
5.356, 5.368 (231.5, 231.0)	D_0	C. IIb. Naturally occurring luminescence from decay of excitons bound to neutral acceptor center. $\hbar\omega$ = 141 meV, S ~ 60. See Fig. 18 and Sec. III.H.
5.482, 5.544 (226.2, 223.6)		A, L', P. IIa, IIb. Intrinsic features of diamond due to creation of free exciton. Energies quoted are thresholds of continuous absorption, Eq. (20a), $\hbar\omega$ = 83 ± 3, 143 ± 2 meV. See Fig. 7 of [140] and Sec. IV.
5.615 (220.8)		A, L', P. IIa, IIb. Intrinsic features of diamond due to creation of free electron hole pair. Energy quoted is threshold of continuous absorption, Eq. (21), $\hbar\omega$ = 143 ± 2 meV. See Fig. 7 of [140] and Sec. IV.

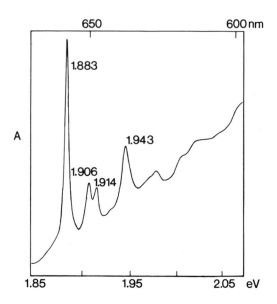

FIG. A1. Absorption spectrum at 80 K of a type Ib synthetic diamond which has been subjected to radiation damage.

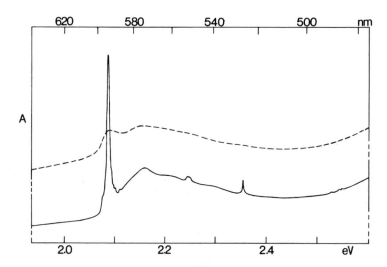

FIG. A2. Absorption in the 2.086-eV band at 80 K (solid line) and 30 K (broken line). The absorption was measured in a natural type Ib diamond after it had been subjected to radiation damage and then annealed at 800 K for a few minutes. Satellite structure at the foot of the zero-phonon line is also observed in type Ia diamond.

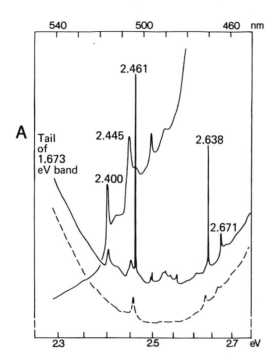

FIG. A3. Absorption near 2.5 eV after radiation damage. Lower
curves are for a type IIa diamond which has been electron irradiated.
Spectra measured at 80 K (solid line) and 300 K (broken line). Curve
rising across the diagram is 80 K absorption of a neutron-irradiated
type IIa diamond which has been annealed at 1100 K for 2 hr. Note
the stability of the 2.400- and 2.445-eV lines, and also their large
widths which are not reduced on cooling to 4K.

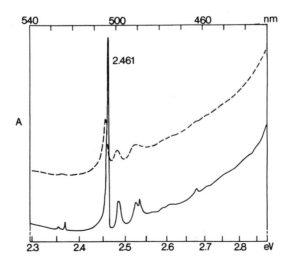

FIG. A4. 2.461-eV absorption band measured at 80 K (solid line)
and 300 K (broken line) after 2-MeV electron irradiation at room
temperature.

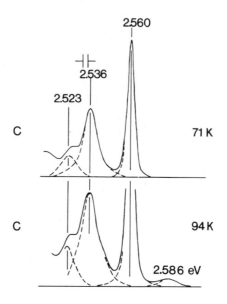

FIG. A5. Cathodoluminescence spectra at 71 and 94 K of a type Ib
synthetic diamond after irradiation with 2 MeV electrons at room temperature. The broken lines give an approximate decomposition, showing
the 2.523- and 2.536-eV components of the emission. These are always
observed with the].560-eV line.

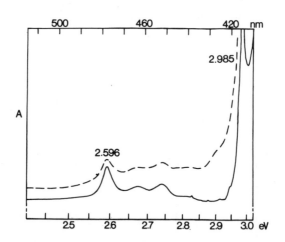

FIG. A6. Absorption spectra at 80 K (solid line) and 300 K
(broken line) of the 2.596-eV (N2) band. The far stronger 2.985-eV
(N3) band is seen to higher energy.

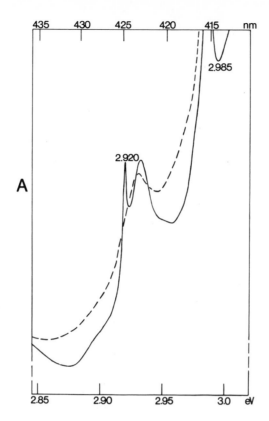

FIG. A7. The 2.920-eV absorption band seen in type IaB diamonds after irradiation with 2-MeV electrons and heating at 900 K
for 2 hr. See also Figs. 35 and 36.

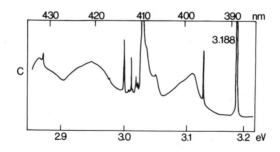

FIG. A8. Detailed cathodoluminescence spectrum at 90 K from a
diamond which has been subjected to electron-radiation damage and
then annealed at 1000 K for 2 hr. The 3.188-eV zero-phonon line is
seen at right; the broad peak at 3.1 eV is the acoustic mode sideband, and the sharper peak near 3.03 eV is the optic mode sideband.
Combination modes appear to lower energy. The sharp lines near 3 eV
apparently always occur with the 3.188-eV line, but the zero-honon
line near 3.13 eV is an independent transition.

127

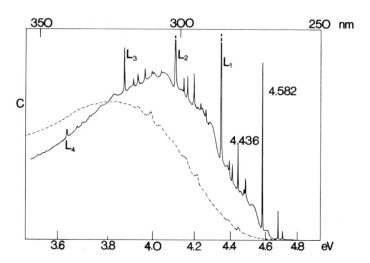

FIG. A9. Catholuminescence from a type II diamond after 2-MeV
electron irradiation, measured at 90 K (solid line) and 300 K (broken
line). Note the unusually large temperature dependence of the energy
of luminescence. Most of the luminescence is from the 4.582-eV
(5RL) band. L_i labels the ith-phonon sideband involving the 237-meV
localized mode. Zero-phonon line at 4.436 eV belongs to a different optical center.

REFERENCES

1. Many myths concerning diamond are summarized by J. R. Sutton,
   in *Diamond a Descriptive Treatise*, Murby, London, 1928. The
   stories listed are quoted by: (a) Sir John Mandeville, (b) Pliny
   the Elder, (c) W. Shakespeare (*A Midsummer Night's Dream*, Act II
   Scene I), (d) Chaucer, (e) Laufer, in (*The Diamond. A Study in
   Chinese and Hellenistic Folk-Lore*, 1915), (f) Joan Evans, in
   (*Magical Jewels*, 1922).

2. H. L. Allsopp, A. J. Burger, and C. van Zyl, *Earth Planet.
   Sci. Lett., 3*, 161 (1967).

3. H. Fesq, D. M. Biddy, C. Erasmus, E. J. Kable, and J. P. F.
   Sellschop, in *Physics and Chemistry of the Earth* (L. H. Ahrens
   et al., Eds.), Pergamon Press, Oxford, 1973.

4. F. A. Raal, *Amer. Mineralogist 42*, 354 (1957); and quoted by
   J. J. Charette, *Physica, 27*, 1061 (1961).

5. W. Kaiser and W. L. Bond, *Phys. Rev., 115*, 857 (1959).

6. E. C. Lightowlers and P. J. Dean, In *Diamond Research*, Industrial Diamond Information Bureau, London, 1964.

7. E. C. Lightowlers, *Anal. Chem.*, *34*, 1398 (1962).

8. (a) E. C. Lightowlers, *Anal. Chem.*, *35*, 1285 (1963). (b) E. C.
    Lightowlers in *Science and Technology of Industrial Diamond*
    (J. Burls, Ed.), Industrial Diamond Information Bureau, London,
    1967, p. 27.

9. G. G. Rocco and O. L. Garzon, *Int. J. App. Radiat. Isotopes*,
    17, 433 (1966).

10. C. E. Melton, C. A. Salotti, and A. A. Giardini, *Amer. Mineral.*,
    57, 1518 (1972).

11. W. A. Runciman and T. Carter, *Solid State Comm.*, *9*, 315 (1971).

12. M. Seal, *Phil. Mag.*, *13*, 645 (1966).

13. F. G. Chesley, *Amer. Mineral.*, *27*, 20 (1942).

14. M. E. Straumanis and E. Z. Aka, *J. Amer. Chem. Soc.*, *73*, 5643
    (1951).

15. E. N. Bunting and A. van Valkenburg, *Amer. Mineral.*, *43*, 102
    (1958).

16. M. S. Freedman, *J. Chem. Phys.*, *20*, 1040 (1952).

17. J. W. Harris, Diamond Conference, Cambridge (1967) unpublished.

18. J. P. F. Sellschop, D. M. Biddy, C. S. Erasmus, and D. W.
    Mingay, in *Diamond Research*, Industrial Diamond Information
    Bureau, London, 1974, p. 43.

19. F. A. Raal, *Proc. Phys. Soc.*, *71*, 846 (1958).

20. M. Lax and E. Burstein, *Phys. Rev.*, *97*, 39 (1955). Their
    formulation of the two-phonon absorption spectrum has been
    used by R. Tubino, L. Piseri, and G. Zerbi, *J. Chem. Phys.*,
    56, 1022 (1972).

21. R. Wehner, M. Borik, W. Kress, A. R. Goodwin, and S. D. Smith,
    Solid State Comm., *5*, 307 (1967).

22. H. M. J. Smith, *Trans. Roy. Soc.*, *241*, A829 (1948).

23. H. B. Dyer, F. A. Raal, L. du Preez, and J. H. N. Loubser,
    Phil. Mag., *11*, 763 (1965).

24. G. Davies, in *Proc. International Conference on Phonons*
    (M. A. Nusimovici, Ed.), Flammarion Sciences, Paris, 1971,
    p. 382.

25. G. B. B. M. Sutherland, D. E. Blackwell, and W. G. Simeral,
    Nature, *174*, 901 (1954).

26. C. D. Clark, in *Physical Properties of Diamond* (R. Berman, Ed.),
    Clarendon Press, Oxford, 1965, p. 297.

27. Yu. A. Klyuev, A. N. Rykov, Yu, A. Dudenkov, and V. M. Zubkov,
    Sov. Phys. Dokl., *14*, 1133 (1970).

28. (a) E. V. Sobolev, V. I. Lisoyvan, and S. V. Lenskaya, *Sov. Phys. Dokl.*, *12*, 665 (1968). (b) P. H. Rainey, Ph.D. thesis, submitted to the University of Reading, U.K., 1974.

29. O. Reinkober, *Ann. Phys.*, *34*, 342 (1911).

30. A. J. Ångström, *Phys. Rev.*, *1*, 597 (1892).

31. W. H. Julius, *Verh. Akad. Weg.*, 1 (1893).

32. W. H. Miller, *Phil. Trans. Roy. Soc.*, *152*, 861 (1862).

33. B. Gudden, and R. W. Pohl, *Z. Phys.*, *3*, 123 (1920).

34. M. Levi, *Trans. Roy. Soc. Can.*, *16*, 243 (1922).

35. F. Peter, *Z. Physik*, *15*, 358 (1923).

36. R. Robertson, J. J. Fox, and A. E. Martin, *Phil. Trans. Roy. Soc.*, *232*, 465 (1934).

37. R. Robertson, J. J. Fox, and A. E. Martin, *Proc. Roy. Soc. London Ser. A*, *157*, 579 (1936).

38. K. G. Ramanathan, *Proc. Ind. Acad. Sci.*, *24*, 130 (1946).

39. K. G. Ramanathan, *Proc. Ind. Acad. Sci.*, *24*, 150 (1946).

40. G. B. B. M. Sutherland and H. A. Willis, *Trans. Farad. Soc.*, *41*, 289 (1945).

41. E. V. Sobolev and V. I. Lisoyvan, *Sov. Phys. Dokl.*, *17*, 425 (1974).

42. E. V. Sobolev, S. V. Lenskaya, and V. I. Lisoyvan, *J. Structural Chem.*, *9*, 917 (1968).

43. T. Evans and C. Phaal, *Proc. Roy. Soc. London Ser. A.*, *270*, 538 (1962).

44. C. E. Wright, Ph.D. thesis, University of Reading, U.K., 1972.

45. G. Davies, *Nature*, *228*, 758 (1970).

46. T. Evans, Diamond Conference, Oxford (1974) unpublished

47. G. Davies and I. Summersgill, in *Diamond Research 1973*, Industrial Diamond Information Bureau, London, 1973, p. 6.

48. C. D. Clark, R. W. Ditchburn, and M. B. Dyer, *Proc. Roy. Soc. London Ser. A.*, *234*, 363 (1956).

49. P. Denham, E. C. Lightowlers, and P. J. Dean, *Phys. Rev.*, *161*, 762 (1967).

50. L. A. Vermeulen and F. R. N. Nabarro, *Phil. Trans. Roy. Soc.*, *A262*, 251 (1967).

51. J. Nahum and A. Halperin, *J. Phys. Chem. Solids*, *24*, 823 (1963).

52. E. A. Konorova, L. A. Sorokin, and S. A. Shevchenko, *Sov. Phys. Solid State*, *7*, 876 (1965).

53. A. G. Redfield, *Phys. Rev.*, *94*, 526 (1954).

54. H. Lenz, *Ann. Phys.*, *77*, 449 (1925).

55. C. L. Klick and R. J. Maurer, *Phys. Rev.*, *81*, 124 (1951).

56. J. F. H. Custers and F. A. Raal, *Nature*, *179*, 268 (1957).

57. R. Mykolajewycz, J. Kalnajs, and A. Smakula, *Appl. Phys. Lett.*, 6, 227 (1965).

58. V. A. Bochko and Yu, L. Orlov, *Sov. Phys. Dokl.*, *15*, 204 (1970).

59. R. Mykolajewycz, J. Kalnajs, and A. Smakula, *J. Appl. Phys.*, *35*, 1773 (1964).

60. T. Evans, Diamond Conference, Bristol (1972) unpublished.

61. S. Caticha-Ellis and W. Cochran, *Acta. Cryst.*, *11*, 245 (1958).

62. P. F. James and T. Evans, *Phil. Mag.*, *11*, 113 (1965). The precise values of the lattice displacements produced by the platelets are unreliable [63]. Annealing studies indicate a monolayer defect structure for the platelets [28b].

63. T. Evans, in *Diamond Research 1973*, Industrial Diamond Information Bureau, London, 1964, p. 2.

64. F. C. Frank, *Proc. Roy. Soc. London Ser. A*, *237*, 168 (1956). See also Ref. 74.

65. C. V. Raman and P. Nilakantan, *Proc. Ind. Akad. Sci.*, *A11*, 389 (1940).

66. K. Lonsdale and H. Smith, *Nature*, *148*, 112 (1941)

67. K. Lonsdale and H. Smith, *Proc. Roy. Soc. London Ser. A*, *179*, 8 (1942).

68. H. J. Grenville-Wells, *Proc. Phys. Soc.*, *65*, 313 (1952).

69. J. A. Hoerni and W. A. Wooster, *Acta. Cryst.*, *8*, 187 (1955).

70. T. Evans and C. E. Wright, Diamond Conference, Cambridge (1971) unpublished.

71. M. Moore and A. R. Lang, *Phil. Mag.*, *25*, 219 (1972).

72. R. J. Elliott, *Proc. Phys. Soc.*, *76*, 787 (1960).

73. A. R. Lang, *Proc. Phys. Soc.*, *84*, 871 (1964).

74. F. C. Frank. *Proc. Phys. Soc.*, *84*, 745 (1964).

75. See also M. Takagi and A. R. Long, *Proc. Roy. Soc. London Ser. A*, *281*, 310 (1964), especially Fig. 5; and Table 1 of [68].

76. V. M. Titova and S. I. Futergendler, *Sov. Phys. Cryst.*, *7*, 749 (1963). The temperatures quoted in this paper are substantially overestimated [44].

77. A. V. Nikitin, M. I. Samoilovich, G. N. Bezrukov, and K. F. Vorozheikin, *Sov. Phys. Dokl.*, *13*, 842 (1969).

78. R. J. Caveney, Diamond Conference, Cambridge (1971) unpublished.

79. L. A. Turk, and P. G. Klemens, *Phys. Rev., B9*, 4422 (1974).

80. M. W. Ackerman and P. G. Klemens, *J. Appl. Phys., 42*, 968 (1971).
     The mass dependence of Eq. (2) was obtained by R. Berman,
     E. L. Foster, and J. M. Ziman, *Proc. Roy. Soc. London Ser. A,
     237*, 344 (1956).

81. D. Gerlich, *J. Chem. Phys. Solids, 35*, 1026 (1974).

82. G. A. Slack, *J. Phys. Chem. Solids, 34*, 321 (1973).

83. B. K. Agrawal and G. S. Verma, *Phys. Rev., 126*, 24 (1962).

84. R. Berman, F. E. Simon, and J. M. Ziman, *Proc. Roy. Soc.
     London Ser. A, 220*, 171 (1953).

85. R. Berman, in *Physical Properties of Diamond* (R. Berman, Ed.),
     Clarendon Press, Oxford, 1965, p. 385.

86. J. J. Charette, *J. Chem. Phys., 35*, 1906 (1961).

87. C. M. Huggins and P. Cannon, *Nature, 194*, 829 (1962).

88. E. V. Sobolev, Yu. A. Litvin, N. D. Samsonenko, V. E. Ill'in,
     S. V. Lenskaya, and V. P. Buruzov, *Sov. Phys. Solid State, 10*,
     1789 (1969).

89. R. M. Chrenko, H. M. Strong, and R. E. Tuft, *Phil. Mag., 23*,
     213-218 (1971).

90. Yu. A. Klyuev, Yu, A. Dudenkov, and V. I. Nepsha, *Geokhimya*,
     1029 (1973).

91. J. F. N. Custers, *Amer. Mineral., 35*, 51 (1950).

92. Y. Kamiya and A. R. Lang, *Phil. Mag., 11*, 347 (1965).

93. E. A. Faulkner, P. W. Whippey, and R. C. Newman, *Phil. Mag.,
     12*, 413 (1965).

94. J. F. Angress and S. D. Smith, *Phil. Mag., 12*, 415 (1965).

95. R. M. Chrenko, R. S. McDonald, and K. A. Darrow, *Nature, 213*,
     474 (1967).

96. W. V. Smith, P. P. Sorokin, I. L. Gelles, and G. J. Lasher,
     Phys. Rev., 115, 1546 (1959).

97. J. H. N. Loubser and L. du Preez, *Brit. J. Appl. Phys., 16*,
     457 (1965).

98. L. A. Shul'man, A. B. Brik, T. A. Nachal'naya, and G. A.
     Podzyarei, *Sov. Phys. Solid State, 12*, 2303 (1971).

99. R. J. Cook and D. H. Whiffen, *Proc. Roy. Soc. London Ser. A,
     295*, 99 (1966).

100. A. B. Lidiard and A. M. Stoneham, in *Science and Technology of
     Industrial Diamonds,* Vol. 1 (J. Burls, Ed.), Eyre and Spottis-
     woode, Grosvenor Press, London, 1967, p. 1.

101. G. D. Watkins and R. P. Messmer, in *Proc. 10th International Conference on the Physics of Semiconductors* (S. P. Keller, J. C. Hensel, and F. Stern, Eds.), United States Atomic Energy Commission, 1970, p. 623; *Phys. Rev. Lett., 25*, 656 (1970).

102. U. Öpik and M. H. L. Pryce, *Proc. Roy. Soc. Lond Ser. A, 238* 425 (1957).

103. M. Caner and R. Englman, *J. Chem. Phys., 44*, 4054 (1966).

104. R. G. Farrer, *Solid State Commun., 7*, 685 (1969).

105. L. A. Shul'man, I. M. Zaritskii, and G. A. Podzyarei, *Sov. Phys. Solid State, 8*, 1842 (1967).

106. J. H. N. Loubser and W. P. van Ryneveld, *Brit. J. Appl. Phys., 18*, 1029 (1967).

107. A. G. Every and D. S. Schonland, *Solid State Comm. 3*, 205 (1965). This calculation is discussed in Sec. 5 of [99].

108. The application of extended Hückel theory to small clusters of atoms as a simulation of crystals has been strongly criticized by F. P. Larkins, *J. Phys., C4*, 3065 (1971); *J. Phys., C4*, 3077 (1971). Suggestions for improved calculations are given by E. B. Moore and C. M. Carlson, *Phys. Rev., B4*, 2063 (1971); see also G. D. Watkins and R. P. Messmer, *Phys. Rev., B4*, 2066 (1971).

109. H. J. Bower and M. C. R. Symons, *Nature, 210*, 1037 (1966).

110. ^{15}N isotope doping produces a shift to lower energy of the 1135, 1100 cm^{-1} peaks to 1120, 1065 cm^{-1}; with no change in the 1350 cm^{-1} localized mode: M. I. Samoilovich, G. N. Bezrukov, V. P. Butozov, and L. D. Podol'skikh, *Dokl. Akad. Nauk. SSSR, 217*, 577 (1974). A theoretical calculation of the one-phonon induced spectrum assuming a T_d nitrogen site was given by A. G. Every, Diamond Conference, Cambridge (1971).

111. P. J. Dean, *Phys. Rev., A139*, 588 (1965).

112. S. Tolansky, in *Diamond Research 1973*, Industrial Diamond Information Bureau, London, 1973, p. 28.

113. R. H. Wentorf and H. P. Bovenkerk, *J. Chem. Phys., 36*, 1987 (1962). Reviews of diamond synthesis are given by R. H. Wentorf in *Advances in High Pressure Research*, Volume 4, edited by R. S. Bradley, Academic Press, New York, 1974, p. 249, and by F. P. Bundy, H. M. Strong, and R. H. Wentorf, Jr., in *Chemistry and Physics of Carbon*, Vol. 10 (P. L. Walker and P. A. Thrower, Eds.), Dekker, New York, 1973, p. 213.

114. R. M. Chrenko, *Nature, 229*, 165 (1971).

115. D. J. Poferl, N. C. Gardner, and J. C. Angus, *J. Appl. Phys., 44*, 1428 (1973); G. N. Bezrukov, V. P. Butuzov, N. N. Gerasimenko, L. V. Lezheiko, Yu. A. Litvin, and L. S. Smitnov, *Sov. Phys. Semicond., 4*, 587 (1970).

116. P. J. Dean, E. C. Lightowlers, and D. R. Wight, *Phys. Rev.,* *A140*, 352 (1965).

117. R. M. Chrenko, *Phys. Rev., B7*, 4560 (1973).

118. S. D. Smith and W. Taylor, *Proc. Phys. Soc., 79*, 1142 (1962).

119. I. G. Austin and R. Wolfe, *Proc. Phys. Soc., B69*, 329 (1956).

120. P. T. Wedepohl, *Proc. Phys. Soc., B70*, 177 (1957).

121. A. T. Collins and A. W. S. Williams, in *Diamond Research 1971*, Industrial Diamond Information Bureau, London, 1971, p. 23; *J. Phys., C4*, 1789 (1971).

122. The semiconducting nature of type IIb diamond is humorously described in the paper announcing their discovery: J. H. F. Custers, *Physica, 18*, 489 (1952). J. E. Brophy, *Phys. Rev., 99*, 1336 (1955) first reported the p-type nature of the semiconduction.

123. W. Kohn, in *Solid State Physics*, Vol. 5 (F. Seitz and D. Turnbull, Eds.), Academic, New York, 1958.

124. C. J. Rauch, in *Proc. International Conference on the Physics of Semiconductors, Exeter, 1962*, Institute of Physics and the Physical Society, London, 1962, p. 276; and *Phys. Rev. Lett., 7*, 83 (1961). See also P. E. Clegg and E. W. J. Mitchell, *Proc. Phys. Soc., 84*, 31 (1964).

125. J. R. Hardy, *Proc. Phys. Soc., 79*, 1154 (1962).

126. J. R. Hardy, *Phil. Mag., 7*, 953 (1962).

127. J. J. Charette, *Physica, 25*, 1303 (1959).

128. A. T. Collins, P. J. Dean, E. C. Lightowlers, and W. F. Sherman, *Phys. Rev., A140*, 1272 (1965).

129. E. O. Kane, *Phys. Rev., 119*, 40 (1960).

130. A. T. Collins, E. C. Lightowlers, and P. J. Dean, *Phys. Rev., 183*, 725 (1969).

131. P. A. Crowther, P. J. Dean, and W. F. Sherman, *Phys. Rev., 154*, 772 (1967).

132. T. Gora, R. Stanley, J. D. Rimstidt, and J. Sharma, *Phys. Rev., B5*, 2309 (1972); J. M. Thomas, E. L. Evans, M. Barker, and P. Swift, *Trans. Farad. Soc., 67*, 1875 (1971).

133. M. Umeno and G. Wiech, *Phys. Stat. Sol., (b)59*, 145 (1973).

134. M. D. Sturge, H. J. Gugenheim, and M. H. L. Pryce, *Phys. Rev., B2*, 2459 (1970).

135. J. R. Hardy, S. D. Smith, and W. Taylor, in *Proc. International Conference on the Physics of Semiconductors, Exeter 1962*, Institute of Physics, London, 1962, p. 521; G. Ruffino and J. J. Charette, *Ricerca Sci., 36*, 526 (1966).

136. A. T. Collins and E. C. Lightowlers, *Phys. Rev.*, *171*, 843 (1968).

137. C. D. Clark, P. Kemmey, E. W. J. Mitchell, and B. W. Henvis, *Phil. Mag.*, *5*, 127 (1960).

138. D. M. S. Bagguley, G. Vella-Coleiro, S. D. Smith, and C. J. Summers, *J. Phys. Soc. Japan*, *21* (Supplement), 244 (1966).

139. E. Anastassakis, *Phys. Rev.*, *186*, 760 (1969).

140. C. D. Clark, P. J. Dean, and P. V. Harris, *Proc. Roy. Soc. London Ser. A*, *277*, 312 (1964).

141. (a) This statement can be roughly justified using the Wilson-Sommerfeld quantization rule for the action of a particle bound in a cyclic state in some coordinate x. $\oint p_x \, dx = hn$, n integral, implies a kinetic energy $k \sim (nh/\ell)^2/2m$ where ℓ is some length characteristic of the extent of the state. Thus for a given potential the exciton will be less strongly bound as $m \to 0$. (b) J. J. Hopfield, in *Physics of Semiconductors, Proc. 7th International Conference* (M Hulin, Ed.), Dunod Editeur, Paris, 1964, p. 725.

142. P. J. Dean, R. A. Faulkner, S. Kimura, and M. Illegems, *Phys. Rev.*, *B4*, 1926 (1971).

143. Being used in this laboratory by E. C. Lightowlers (1974).

144. G. H. Glover, *Solid State Electron.*, *16*, 973 (1973).

145. E. C. Lightowlers and A. T. Collins, *Phys. Rev.*, *151*, 685 (1966).

146. A. W. S. Williams, E. C. Lightowlers, and A. T. Collins, in *Diamond Research 1970*, Industrial Diamond Information Bureau, London, 1970, p. 23.

147. D. R. Wight and P. J. Dean, *Phys. Rev.*, *154*, 689 (1967).

148. P. Denham and E. C. Lightowlers, *Phys. Rev.*, *174*, 800 (1968).

149. A. W. S. Williams, E. C. Lightowlers, and A. T. Collins, *J. Phys.*, *C3*, 1727 (1970).

150. Doubts about the identification were first published by P. T. Wedepohl, *J. Phys.*, *C1*, 1773 (1968).

151. E. C. Lightowlers, A. T. Collins, P. Denham, and P. S. Walsh, in *Diamond Research 1968*, Industrial Diamond Information Bureau, London, 1968, p. 11.

152. C. D. Clark, R. W. Ditchburn, and H. B. Dyer, *Proc. Roy. Soc. London Ser. A*, *237*, 75 (1956).

153. C. Phaal, *Phil. Mag.*, *11*, 369 (1965).

154. J. L. Warren, J. L. Yarnell, G. Dolling, and R. A. Cowley, *Phys. Rev.*, *158*, 805 (1967). See also J. L. Yarnell, J. L. Warren, and R. G. Wenzel, *Phys. Rev. Lett.*, *13*, 13 (1964).

155. G. Peckham, *Solid State Comm.*, *5*, 311 (1967).

156. G. Dolling and R. A. Cowley, *Proc. Phys. Soc., 88*, 463 (1966).

157. R. Wehner, *Phys. Stat. Sol., 17*, K179 (1966).

158. S. A. Solin and A. K. Ramdas, *Phys. Rev., B1*, 1687 (1970).

159. J. R. Hardy and S. D. Smith, *Phil. Mag., 6*, 1163 (1961).

160. F. A. Johnson and R. Loudon, *Proc. Roy. Soc. London Ser. A, 281*, 274 (1964).

161. H. Bilz, R. Geick, and K. F. Renk, in *Lattice Dynamics*, Pergamon, Oxford, 1965, p. 355.

162. E. Burstein and S. Ganeson, *J. Physique, 26*, 637 (1965).

163. J. F. Angress, C. Cooke, and A. J. Maiden, *J. Phys., C1*, 1769 (1968). Similar results have been given by E. Anastassakis, S. Iwasa, and E. Burstein, *Phys. Rev. Lett., 17*, 1051 (1966) except that their spectrometer calibration seems to be in error.

164. J. F. Angress and A. J. Maiden, *J. Phys., C4*, 235 (1971).

165. R. S. Krishnam, *Proc. Ind. Acad. Sci., A26*, 399 (1947).

166. C. V. Raman, *Proc. Ind. Acad. Sci., A44*, 99 (1956).

167. D. Krishnamurti, *Proc. Ind. Acad. Sci., A40*, 211 (1954).

168. A. Mani, *Proc. Ind. Acad. Sci., A20*, 117 (1944).

169. C. Ramaswamy, *Nature, 125*, 704 (1930).

170. P. S. Narayanan, *Proc. Ind. Acad. Sci., A32*, 1 (1950).

171. R. S. Krishnan, *Proc. Ind. Acad. Sci., A24*, 45 (1946).

172. P. G. N. Nayar, *Proc. Ind. Acad. Sci., 13*, 284 (1941).

173. S. S. Mitra, O. Brafman, W. B. Daniels, and R. K. Crawford, *Phys. Rev., 186*, 942 (1969).

174. W. J. Borer, S. S. Mitra, and K. V. Namjoshi, *Solid State Comm., 9*, 1377 (1971).

175. A. K. McQuillan, W. R. L. Clements, and B. P. Stoicheff, *Phys. Rev., A1*, 628 (1970).

176. F. Stenman, *J. Appl. Phys., 40*, 4164 (1969).

177. A. Laubereau, D. von der Linde, and W. Kaiser, *Phys. Rev. Lett., 27*, 802 (1971).

178. M. Born and K. Huang, *Dynamical Theory of Crystal Lattices*, Clarendon, Press, Oxford (1954).

179. M. H. Cohen and J. Ruvald, *Phys. Rev. Lett., 23*, 1378 (1969); C. H. Wu and J. L. Birman, *J. Phys. Chem. Solids, 36*, 305 (1975).

180. B. J. Parsons and C. D. Clark, Diamond Conference, Reading (1973) unpublished; *Proc. R. Soc.* (1976) in press.

181. C. D. Clark, *J. Phys. Chem. Solids, 8*, 481, (1959).

182. P. J. Dean and J. C. Male, *Proc. Roy. Soc. London Ser. A, 277*, 330 (1964).

183. P. J. Dean and I. H. Jones, *Phys. Rev., A133*, 1698 (1964).

184. G. Davies, *J. Phys. Chem. Solids, 31*, 883 (1970).

185. Eq. 328 of R. A. Smith, in *Semiconductors*, University Press, Cambridge (1964).

186. H. R. Philip and E. A. Taft, *Phys. Rev., 127*, 159 (1962).

187. W. C. Walker and J. Osantowski, *Phys. Rev., A134*, 153 (1964).

188. H. R. Philip and E. A. Taft, *Phys. Rev., A136*, 1445 (1964).

189. R. A. Roberts and W. C. Walker, *Phys. Rev., 161*, 730 (1967).

190. R. A. Roberts, D. M. Roessler, and W. C. Walker, *Phys. Rev. Lett., 17*, 302 (1966).

191. L. R. Savaria and D. Brust, *Phys. Rev., 170*683 (1968).

192. N. R. Whetton, *Appl. Phys. Lett., 8*, 135 (1966).

193. The Ivey-Mollwo dependence of F-center energies on alkali halide lattice parameter can be quoted in support of this over-simplified model: W. Gebhardt and H. Kühnert, *Phys. Stat. Sol., 14*, 157 (1966).

194. T. H. Keil, *Phys. Rev., A140*, 601 (1965).

195. A. A. Maradudin, *Solid State Phys., 18*, 273 (1966).

196. M. H. L. Pryce, in *Phonons in Perfect Lattices and in Lattices with Point Imperfections* (R. W. H. Stevenson, Ed.), Oliver and Boyd, London, 1966, p. 403.

197. D. B. Fitchen, in *Physics of Color Centers* (W. B. Fowler, Ed.), Academic, New York, 1968, p. 293.

198. M. Mostoller, B. Henderson, W. A. Sibley, and R. F. Wood, *Phys. Rev., B4*, 2667 (1971).

199. P. Giesecke, W. von der Osten, and N. Röder, *Phys. Stat. Sol., 51*, 723 (1972).

200. A. Halperin and O. Nawi, *J. Phys. Chem. Solids, 28*, 2175 (1967).

201. G. Davies, *J. Phys., C7*, 3797 (1974).

202. P. A. Crowther and P. J. Dean, *J. Chem. Phys. Solids, 28*, 1115 (1967).

203. R. Wedlake, Ph.D. thesis, University of Reading, U.K., 1970.

204. G. Davies and E. C. Lightowlers, *J. Phys., C3*, 638 (1970).

205. H. B. Dyer and L. duPreez, *J. Chem. Phys., 42*, 1898 (1965).

206. H. B. Dyer and L. du Preez, in *Science and Technology of Industrial Diamonds,* Vol. 1 (J. Burls, Ed.), Industrial Diamond Information Bureau, London, 1967, p. 23.

207. L. du Preez, Ph.D. thesis, University of Witwatersrand, South Africa (1965).

208. In silicon see, e.g., G. D. Watkins, in *Radiation Damage in Semiconductors* (G. D. Watkins, Ed.) Academic Press, New York, 1965, p. 97, and J. W. Corbett, in *Electron Radiation Damage in Semiconductors and Metals, Solid State Physics Supplement 7* (F. Seitz and D. Turnbull, Eds.), Academic Press, New York, 1966, p. 78. In germanium see A. Kiraki, J. W. Cleland, and J. H. Crawflrd, *J. Appl. Phys.*, *38*, 3519 (1967), *Phys. Rev.*, *177*, 1203 (1969).

209. F. S. Ham, *Phys. Rev.*, *A138*, 1727 (1965).

210. The expression differs for S_E by a factor of four from that given by F. S. Ham, *Phys. Rev.*, *166*, 307 (1968); this is because of the different definitions of the $E\theta$ and $E\varepsilon$ strains.

211. M. C. M. O'Brien, *J. Phys.*, *C4*, 2408 (1971).

212. J. Duran, Y. M. d'Aubigne, and R. Romestain, *J. Phys.*, *C5*, 2225 (1972) have given S_T for six-fold coordination.

213. R. S. Krishnan, *Proc. Ind. Acad. Sci.*, *A24*, 33 (1946); J. Thewlis and A. R. Davey, *Phil. Mag.*, *1*, 408 (1956); B. J. Skinner, *Amer. Mineral.*, *42*, 39 (1957); S. I. Novikova, *Sov. Phys. Solid State*, *2*, 1464 (1961).

214. H. J. McSkimmin, P. Andreatch, Jr., and P. Glynn, *J. Appl. Phys.*, *43*, 985 (1972) give $c_{11} = (10.79 \pm 0.05) \times 10^{12}$, $c_{12} = (1.24 \pm 0.05) \times 10^{12}$, $c_{44} = (5.78 \pm 0.02) \times 10^{12}$ dyn cm^{-2} at 25°C. Other references are given in the paper.

215. B. Walter, *Ann. Phys. Chem.*, *42*, 505 (1891).

216. P. G. N. Nayar, *Proc. Ind. Acad. Sci.*, *A14*, 1 (1941).

217. W. A. Runciman, *Proc. Phys. Soc.*, *86*, 629 (1965).

218. A. A. Kaplyanskii, V. I. Kolyshkin, and V. N. Medvedev, *Sov. Phys. Solid State*, *12*, 1193 (1970).

219. C. D. Clark and C. A. Norris, *J. Phys.*, *C3*, 651 (1970); R. J. Elliott, I. G. Matthews, and E. W. J. Mitchell, *Phil. Mag.*, *3*, 360 (1958).

220. This splitting of 0.59 meV is not to be confused with the much larger splittings of 1.5 to 4.3 meV reported by A Mani, *Proc. Ind. Acad. Sci.*, *A19*, 231 (1944). This was probably merely self-absorption of the emission of the 2.985-eV line.

221. J. H. N. Loubser and A. C. J. Wright, in *Diamond Research 1973*, Industrial Diamond Information Bureau, London, 1973, p. 16.

222. E. V. Sobolev, *Geol. Geofiz.*, *12*, 127 (1969).

223. E. W. J. Mitchell, in *Diamond Research 1973*, Industrial Diamond Information Bureau, London, 1963.

224. A. E. Hughes and W. A. Runciman, *Proc. Phys. Soc.*, *90*, 827 (1967). These workers used a basic trigonal center which is not compatible with the diamond structure since a [$\bar{1}$ 1 0] axis cannot lie in a reflection plane of [1 1 1]-oriented trigonal center. However their final expressions are still applicable.

225. H. B. Dyer and P. Ferdinando, *Brit. J. App. Phys.*, *17*, 419 (1966).

226. A. T. Collins, Diamond Conference, Oxford (1974) unpublished; *Proceedings of the International Conference on Radiation Damage Effects*, Dubrovnik (1976) edited by M. A. Urli, Institute of Physics (to be published).

227. J. Walker, L. A. Vermeulen, andC. D. Clark, *Proc. Roy. Soc. London Ser. A, 341*, 253 (1974).

228. R. Farrar and L. A. Vermeulen, Diamond Conference, Oxford (1974) unpublished.

229. C. D. Clark and C. A. Norris, *J. Phys.*, *C4*, 2223 (1971).

230. C. D. Clark and J. Walker, *Diamond Research 1972*, Industrial Diamond Information Bureau, London, 1972, p. 2.

231. G. Davies, *Proc. Roy. Soc.*, *A336*, 507 (1974).

232. M. Crossfield, private communication (1974).

233. C. D. Clark and J. Walker, *Proc. Roy. Soc. London Ser. A, 234*, 241 (1973).

234. (a) G. Davies and C. M. Penchina, *Proc. Roy. Soc. London Ser A, 338*, 359 (1974). (b) G. Davies, *J. Phys.*, *C8*, 2448 (1975).

235. D. R. Wight, P. J. Dean, E. C. Lightowlers, and C. D. Mobsby, *J. Luminescence, 4*, 169 (1971).

236. S. A. Solin, *Phys. Lett.*, *38A*, 101 (1972).

237. F. P. Larkins and A. M. Stoneham, *J. Phys.*, *C4*, 143 (1971).

238. Early calculations by Lidiard and Stoneham gave w_E = 149 meV, w_T = 105 meV [100]. Later work gave w_A = 63 meV, w_T = 75 meV, w_T = 65 meV [237].

239. M. A. Stoneham, Diamond Conference, Oxford (1974).

240. The sequence of events as higher terms in the coupling are introduced is described by R. Englman in *The Jahn-Teller effect in Molecules and Crystals,* Wiley-Interscience, London, 1972, p. 20.

241. M. C. M. O'Brien, *Proc. Roy. Soc. London Ser. A, 281*, 323 (1964), Eqs. (17) and (23).

242. F. S. Ham, in *Electron Paramagnetic Resonance* (S. Geshwind, Ed.), Plenum Press, New York, 1972, p. 1.

243. The signs of e given in [234] are based on the sign of the electric dipole matrix element being the same for the vibronic A to T transition as for an electronic A to T transition. The sign in fact is undetermined at present since it depends on the reason why the A to T transition is allowed.

244. H. C. Longuet-Higgins, U Öpik, M. H. L. Pryce, and R. A. Sack, *Proc. Roy. Soc. London Ser. A, 244*, 1 (1958).

245. The clarity with which this combination mode would be seen is illustrated by Fig. 1 of J. T. Ritter, *Solid State Comm.*, *8*, 773 (1970).

246. D. S. Nedvetskii and V. A. Gaisin, *Sov. Phys. Solid State*, *14*, 2533 (1973).

247. R. G. Farrar and L. A. Vermeulen, *J. Phys.*, *C5*, 2762 (1972).

248. M. Lannoo and A. M. Stoneham, *J. Phys. Chem. Solids*, *29*, 1987 (1968).

249. Recent reviews of theoretical work are: A. M. Stoneham, in *Radiation Effects in Semiconductors*, Gordon and Breach, London, 1971, p. 7, and A. B. Lidiard, in *Radiation Damage and Defects in Semiconductors*, Institute of Physics, London, 1972, p. 238. F. P. Larkins has discussed one of Lidiard's suggestions in *J. Phys.*, *C6*, L345 (1973). A discussion of the difficulties in formulating a satisfactory approach is given by C. A. Coulson, in *Radiation Damage and Defects in Semiconductors*, Institute of Physics, London, 1972, p. 249. Extended Hückel calculations are given by R. P. Messmer and G. D. Watkins, *Rad. Effects*, *9*, 9 (1971); see also G. D. Watkins and R. P. Messmer, *Phys. Rev. Lett.*, *32*, 1244 (1974). Other papers not separately listed include: J. Friedel, M. Lannov, and G. Leman, *Phys. Rev.*, *164*, 1056 (1967); C. A. Coulson and F. P. Larkins, *J. Chem. Phys. Solids*, *32*, 2245 (1971); F. P. Larkins, *J. Chem. Phys. Solids*, *32*, 2123 (1971); F. P. Larkins and A. M. Stoneham, *J. Phys.*, *C4*, 154 (1971).

250. M. F. Hamer and G. Davies, unpublished (1974).

251. H. R. Fetterman and D. B. Fitchen, *Solid State Comm.*, *6*, 501 (1968).

252. D. M. Dennison and G. E. Uhlenbeck, *Phys. Rev.*, *41*, 313 (1932).

253. See, for example, D. Bohm, *Quantum Theory*, Prentice-Hall, London, 1960, p. 281.

254. A pair of parabolae is required in the electronic ground state so that there is a wavefunction of even parity and a wavefunction of odd parity associated with each energy level. If there were a single ground state parabola, the wavefunction of each level would have an even *or* odd parity. The one-phonon sideband would then appear as a single line in both absorption and emission.

255. G. Davies, *J. Phys.*, *C5*, 2534 (1972).

256. (a) M. F. Hamer and (b) E. de Sa, private communicaton (1975).

257. A. A. Kaplyanskii, *Opt. Spectros.*, *16*, 329 (1964).

258. D. S. Nedvetskii and V. A. Gaisin, *Sov. Phys. Solid State*, *16*, 145 (1974).

259. R. K. Wild, T. Evans, and A. R. Lang, *Phil. Mag.*, *15*, 267 (1967).

260. M. A. Stoneham, *Rev. Mod. Phys.*, *41*, 82 (1969).

261. J. D. Eshelby, *Solid State Phys.*, *3*, 79 (1956).

262. G. Davies and M. Crossfield, *J. Phys.*, *C6*, L104 (1973).

263. G. Davies, *J. Phys.*, *C3*, 2474 (1970).

264. Chaumet reported inexplicable differences in the luminescence
     of diamonds in *Compt. Rend.*, *134*, 1139 (1902).

265. E. V. Sobolev and V. N. Krasnitsa, *Dokl. Akad. Nauk. SSSR*,
     212, 709 (1973).

266. M. Crossfield, G. Davies, A. T. Collins, and E. C. Lightowlers,
     J. Phys., *C7*, 1909 (1974).

267. M. F. Thomaz, M. F. Thomaz, and C. L. Braga, *J. Phys.*, *C5*, L1
     (1972) reported a value of 23 ± 4 nsec for the 2.463-eV radia-
     tive lifetime. This was obtained by photoluminescence, with
     excitation between 300 and 400 nm, and 2.985-eV (N3) lumines-
     cence was also excited, its lifetime being 40 nsec. Excita-
     tion with lower energy photons later gave T_r = 16 ± 2 nsec
     for the 2.463-eV lifetime, M. F. Thomaz, private communication
     (1973).

268. D. L. Dexter, *J. Chem. Phys.*, *21*, 836 (1953).

269. Chaumet, *Compt. Rend.*, *134*, 1139 (1902), records his horror at
     a particularly valuable diamond he had borrowed being turned
     brown for several hours by exposure to mercury light.

270. P. Pringsheim, *Phys. Rev.*, *91*, 551 (1953).

271. R. G. Farrar and L. A. Vermeulen, *J. Phys.*, *C5*, 2762 (1972).

272. The simple saturating exponential growth is a consequence of
     using type Ib diamonds. The large 3.150-ev band is almost
     unchanged in strength in producing large fractional changes
     of the weak 1.673-eV strength. The ultraviolet continuum asso-
     ciated with the 1.673-eV center is very weak in comparison with
     the 3.150-eV band so that there is negligible competition for
     the light.

273. D. G. Thomas, *J. Phys.*, *D2*, 637 (1969).

274. Formal treatments of the theory have been given by F. E.
     Williams, *J. Phys. Chem. Solids*, *12*, 265 (1960).

275. P. J. Dean, C. H. Henry, and C. J. Frosch, *Phys. Rev.*, *168*,
     812 (1968).

276. D. G. Thomas, J. J. Hopfield, and W. M. Augustynick, *Phys.
     Rev.*, *A140*, 202 (1965).

277. E. Zacks and A. Halperin, *Phys. Rev.*, *B6*, 3072 (1972).

278. P. J. Dean and J. C. Male, *J. Phys. Chem. Solids*, *25*, 1369 (1964).

279. G. N. Bezrukov, V. P. Butuzov, N. N. Gerasimenko, L. V. Lezheiko,
     Yu, A. Litvin, and L. S. Smirnov, *Sov. Phys. Semicond.*, *4*, 587
     (1970).

280. G. Davies, *Diamond Research*, Industrial Diamond Information
      Bureau, London, 1975, p. 13.

281. P. S. Walsh, E. C. Lightowlers, and A. T. Collins, *J.
      Luminescence, 4*, 369 (1971).

282. R. Chen, S. A. A. Winer, and N. Kristianpoller, *J. Chem. Phys.,
      60*, 4804 (1974).

283. Recent papers include M. Ya. Shcherbakova, E. V. Sobolev,
      N. D. Samsonenko, and V. K. Aksenkov, *Sov. Phys. Solid State,
      11*, 1104 (1969); J. H. N. Loubser and A. C. J. Wright, *J.
      Phys., D6*, 1129 (1973); M. Ya. Shcherbakova, E. V. Sobolev,
      N. D. Samsonenko, V. A. Nadolinnyi, P. V. Schastnev, and
      A. G. Semenov, *Sov. Phys. Solid State, 13*, 281 (1971); J. N.
      Laner and A. M. A. Wild, *Phil. Mag., 24*, 274 (1971); Y. M. Kim
      and G. D. Watkins, *J. App. Phys., 42*, 722 (1971); Yu. A.
      Bratashevskii, F. N. Bukhan'ko, N. D. Samsonenko, and O. Z.
      Shapiro, *Sov. Phys. Solid State, 13* , 1809 (1972); L. A.
      Shul'man, A. B. Brik, T. A. Nachal'naya, and G. A. Podzyarei,
      Sov. Phys. Solid State, 12, 2303 (1971).

284. J. Owen, in *Physical Properties of Diamond* (R. Berman, Ed.),
      Clarendon Press, Oxford, 1965, p. 274.

285. A. Halperin and J. Nahum, *J. Phys. Chem. Solids, 18*, 297 (1961);
      A. Halperin and R. Chen, *Phys. Rev., 148*, 839 (1966).

286. First made by Andrew Pritchard in London during 1824 to the
      order of Sir David Brewster. The lenses were made obsoles-
      cent in about 1840 when cheaper compound lenses became avail-
      able. See R. Nuttall and A. Frank, *New Sci.*, 92 (13 January
      1972) and S. Tolansky in *Science and Technology of Diamond*,
      Vol. 2 (J. Burls, Ed.), Eyre and Spottiswoode, Grosvenor Press,
      London, 1967, p. 341.

287. R. Boyle, in *An Essay about the Origins and Virtues of Gems*,
      London, 1962.

288. Albertus Magnus, in *De Rebus Metallicis et Mineralibus*, quoted
      by J. W. Mellor, *A Comprehensive Treatise on Inorganic and
      Theoretical Chemistry*, Longmans, Green, London, 1922.

289. I. Newton, in *Optiks*, 4th edition reprinted by G. Bell and Son
      Ltd., London, 1931, pp. 272-274.

290. W. Crookes, *Phil. Trans. Roy. Soc., 170*, 135; *170*, 641 (1879).

291. For example, A. Miethe, *Ann. Physik, 19*, 633 (1906).

292. D. R. Wight, Ph.D. thesis, University of London, 1968.

293. J. E. Ralph, *Proc. Phys. Soc., 76*, 688 (1960).

294. B. Burton, Diamond Conference, Bristol (1972).

295. E. V. Sobolev, V. E. Il'in, and O. P. Yur'eva, *Sov. Phys.
      Solid State, 11*, 938 (1969).

296. J. Walker, "Optical and uniaxial stress studies of irradiation-
      induced defects in diamond," *Proceedings of the International
      Conference on Lattice Defects in Semiconductors*, Freiburg,
      1974, edited by F. A. Huntley, Institute of Physics, London,
      1975, p. 317.

297. J. F. Angress, A. R. Goodwin, and S. D. Smith, *Proc. Roy. Soc.
      London Ser. A, 308*, 111 (1968).

FRACTURE IN POLYCRYSTALLINE GRAPHITE

J. E. Brocklehurst

United Kingdom Atomic Energy Authority
Reactor Fuel Element Laboratories
Springfields, Salwick,
Preston, Lancashire,
England

I.	INTRODUCTION	146
II.	MATERIAL	147
III.	DEFORMATION AND ASSOCIATED CHARACTERISTICS	150
	A. Deformation.	150
	B. Acoustic Emission.	153
	C. Microscopic Examination.	154
	D. Effects of Prestress on Young's Modulus and Strength	157
	E. Micromechanisms of Failure	160
IV.	FRACTURE UNDER UNIFORM STRESS.	165
	A. The Griffith Model	166
	B. Uniaxial Strength.	169
	C. Biaxial Strength	170
	D. Triaxial Strength.	176
V.	STATISTICS OF FRACTURE, SIZE EFFECTS, AND NONUNIFORM STRESS CONDITIONS	180
VI.	EFFECTIVE WORK OF FRACTURE, FRACTURE TOUGHNESS, AND INHERENT DEFECT SIZE	193
VII.	FATIGUE.	204
VIII.	NOTCH SENSITIVITY.	218
IX.	EFFECT OF DENSITY, GRAIN SIZE, AND CRYSTALLINITY	225
X.	EFFECT OF TEMPERATURE, ATMOSPHERE, AND STRAIN RATE	231
	A. Temperature and Atmosphere	231
	B. Temperature and Strain Rate.	236
	C. Impact and Thermal Shock	240
XI.	EFFECT OF FAST-NEUTRON IRRADIATION	243
XII.	EFFECT OF INTERCALATION.	258

XIII. SUMMARIZING DISCUSSION 264
       REFERENCES 272

I. INTRODUCTION

One of the outstanding technological problems in engineering designs
incorporating graphite components is an adequate quantitative description of the failure criteria under general states of loading. The
design engineer requiring knowledge of the limiting conditions gov-
erning the failure of a component in a given environment and stressed
state, both of which are often time dependent, rightly criticizes
the materials scientist who is only able to define the failure state
under particular conditions, and whose qualitative understanding of
the fracture process at the microscopic level is not sufficiently
developed quantitatively to enable him to define the macroscopic
criteria for failure in more general situations.

Failure criteria are usually phenomenological descriptions of
fracture derived from laboratory tests and expressed in terms of the
limiting states of stress and/or strain assuming the material is
homogeneous on the macroscopic scale. Ideally, any postulated criterion to be applied to structures in service should also adequately
describe and interrelate the results of different types of laboratory
fracture tests. At the microscopic level, the graphite structure
is complex and far from homogeneous. A desirable objective is a
detailed and quantitative mechanistic explanation of crack initiation and growth, together with the conditions leading to a macroscopic crack and ultimate failure. This description must be linked
to the externally imposed stress and environmental conditions on the
material, which is normally treated as a continuum in stress analysis
problems. There is still, however, a wide gap between physically
based models and the phenomenological theories relevant to the
engineer.

Jenkins [1] has recently described the deformation characteristics of carbons and graphites, and the aim of this chapter is to

concentrate on the fracture conditions, although a full discussion
of this topic is not possible without some consideration of the
deformation behavior and its associated characteristics. Over the
years, specific aspects of graphite fracture have been studied on
a wide range of materials, and Jouquet [2] has recently given a
brief summary of some of this work. Earlier reviews include those
of Losty [3], Rappeneau and Jouquet [4], and Reynolds [5]. The pur-
pose of the present review is to bring together these different
aspects of graphite fracture and give a detailed and comprehensive,
if not completely exhaustive, survey of the current position. An
overall picture must unavoidably involve some generalizations because of the many different materials studied, but the aim is to
clarify some of the remaining problems and to identify gaps in existing knowledge. It should also provide a starting point for new
researchers in this area.

II. MATERIAL

This review is concerned mainly with studies on well-graphitized
polycrystalline materials. They cover a range of materials manu-
factured from filler coke particles, usually formed with a binder
and impregnated to increase density, although the process may differ
in some speciality graphites. The materials range widely in filler
particle type and size, binder type, and pore structure, all of
which strongly influence the properties of the material. Graphite
crystals are highly anisotropic, and the extent of anisotropy in
the polycrystal is dictated both by the degree of crystallite orientation within the filler particle or grain and by the particle orientation caused by its geometric shape responding to the forming process. For example, the filler grains may be needle-like with a highly
oriented structure and a preferred deformation by slip along the
layers, or they may be more spherical in shape and deform less easily.
The matrix binder or impregnant has generally a more random crystallite structure. Most graphites are described as orthotropic with
one principal material direction corresponding to the forming

direction; the other two directions lie on a plane normal to this
direction. Thus any anisotropy in material properties occurs with
the major difference between those measured parallel to the forming
direction and those measured perpendicular, extrusion and molding
producing differences in an opposite sense.

Typical densities of many commercially useful materials range
from 1.7 to 1.9 g/cm^3 with some 20% total porosity when compared with
the theoretical value of 2.26 g/cm^3. The material contains a wide
spectrum of pores, from small microcracks to large voids, which
arise from a variety of sources including differential thermal contraction at misoriented crystallite and grain boundaries, blow
holes due to evolution of volatiles during manufacture, or inherent
voids in the starting cokes. It is convenient for some purposes to
visualize the materials as containing at one end of the scale a
number of large pores, which determine, for example, the internal
stress distribution for an externally applied load, and at the other,
small intercrystalline microcracks, which, for example, influence
the crystal contributions to the bulk properties.

These microcracks, postulated by Mrozowski [6] to arise as a
result of the restraints on anisotropic crystal contraction from the
graphitizing temperature and to be associated with a "frozen-in"
residual stress pattern, have been observed in electron microscope
studies [7] of certain graphites to be ∿10^{-2} mm, oriented parallel
to the crystallite basal planes. The remaining pores range widely
in size and typically have maximum dimensions similar to or greater
than those of the filler particles, which in some materials are
approximate millimeters. They exist either as isolated voids or as
continuous porosity. Smith [8] and Pears [9] describe this porosity
in two distinct categories: (i) the many relatively small background
defects which are uniformly dispersed and are always present; and
(ii) the fewer large "disparate" voids which could, in general,
be avoided in manufacture and, for exmaple, can be aligned to give
a combined effect which may be very damaging to the structure.

Table 1 lists some of the polycrystalline graphites discussed
in this review, with property data either from Mantell [10] or from

TABLE 1

Some of the Graphites Discussed in this Review[a]

Grade	Source	Forming method (Extrusion) (Molding)	Bulk density (g/cm³)	Young's modulus (GN/m²) parallel/perpendicular to forming axis	Strength (MN/m²) parallel or parallel/perpendicular to forming axis		
					Tensile	Bend	Compressive
PGA	U.K.	E	1.7	11/6	10/6	14/9	30/30
IM1-24		M	1.8	10/10	22	33	85
SM2-24		M	1.7	8.0/8.5	12	19	47
VNMC		M	1.75	8.5/9	11	19	51
AGOT	U.S.A.	E	1.7	10/8	10/9	16/13	41/41
AUC		E	1.7	11/6	7	15	31
ATJ		M	1.75	8/10	10/12	25/28	59/57
ATJ-S		M	1.85	9/13	30		90
H205		M	1.75	9.5/11		28	70
RC-4		E		10/	12	23	
Graphitite G		E	1.9	10/7	23	38	78
AXF-Q1		M	1.85	10/10	33	96	138
ZTA		Hot pressed	1.95	6/18	8/28	17/37	80/50
MPG-6	U.S.S.R.	M	1.8	10/	32	52	98
Glassy carbon					200		1400
Carbon fibers				200	1000		

[a]Data from literature sources in the text or from reference 10.

the various literature sources referred to in the text. While some
of the data may not be fully typical of the average product, they
serve to show the variation in strength properties of the different
materials with some indication of the degree of anisotropy. Young's
modulus data are given parallel and perpendicular to the forming
direction. Whether this direction represents "with grain" or "against
grain" directions depends on whether extrusion or molding is used.
Strength data in Table 1 are in general given parallel to forming,
although both directions are quoted for some of the more anisotropic
graphites. Data for glassy carbon and carbon fibers are included
to show the high strength obtained in these materials due to the high
contribution of the strong covalent C-C bond and a minimum number of
weakening defects.

The mechanism of the fracture process in polycrystalline graphite can be considered at three structural levels: the atomic dislocations, the microcracks, and the gross macropores. The engineer is
concerned with the limiting stress-strain states of a particular
graphite in the macro sense, and the materials scientist is concerned
with interrelating the mechanisms at these different levels to give
a full description of the fracture process. It is necessary first,
however, to consider the events occurring prior to ultimate failure.

III.   DEFORMATION AND ASSOCIATED CHARACTERISTICS

A.   *Deformation*

Jenkins [1] has reviewed the deformation behavior of carbons and
graphites in a recent contribution to this series, but it is necessary to any discussion on fracture to consider the events prior
to ultimate failure. It is hoped that this section will provide
some continuity with Jenkins' review.

The deformation characteristics of polycrystalline graphite have
been studied in detail by several authors, notably Losty and Orchard
[11], Seldin [12] and Greenstreet et al. [13], in uniaxial stress
modes, and also, as seen later, by other workers under multiaxial
stresses. The general behavior is a nonlinear stress-strain

relationship, a hysteresis effect on unload-reload curves with a
permanent set at zero load, which can be recovered on thermal annealing, and a decreasing elastic modulus with increasing stress. This
behavior was described by Jenkins [14] originally on a rheological
model in which an increasing proportion of the elements deform plas-
tically against the restraining effect of an elastic matrix. This
model gave a parabolic relationship which deviated from the observed
behavior at high compressive stress levels, and Jenkins [15] developed an analysis based on a dislocation model of the spread in plastic yield from the tips of the pre-existing microcracks, the matrix
remaining elastic.

The theme of Jenkins' recent review [1] is that the response of
any carbon to an applied stress is governed by a continuous carbon
network, providing an elastic boundary restraint to the shear deformation of component grains, and accounting for the "memory" displayed
by graphite and carbon bodies for their original state. Mechanical
hysteresis and the apparent permanent deformation observed at low
temperatures are attributed to internal friction associated with
interlamellar shear, and the material is, basically, wholly elastic.
This basic model allows the description of a wide range of carbon
materials from those whose deformation is dominated by the restraining carbon-carbon network, e.g., glassy carbons, to those more perfect graphite structures, containing large regions of near-perfect
basal planes, in which boundary restraint is relatively small.

This behavioral model is not necessarily in conflict with the
view that in well-graphitized materials deformation is dominated by
basal plane shear. Simmons [16] and Davidson and Losty [17] were
among the first to suggest that elastic and plastic deformation in
graphite was accommodated by the basal plane shear and slip of crystallites with the layer planes moving over each other like cards in
a pack. Glissile dislocations within the basal plane providing a
mechanism for such slip were identified, for example, by Amelinckx
and Delavignette [18], Williamson [19], and later by Baker and Kelly
[20], and this model has been the basis for discussions of the elastic
properties of graphite, for example, by Kelly [21] and Reynolds [22]

and also for interpretation of irradiation induced creep of graphite
[23].   Where Jenkins [1] does take issue with other interpretations
is on the magnitude of the contribution made by the movement of these
dislocations to the high shear compliance of the graphite crystal.
He argues that dislocations only contribute about 10% to the crystal
shear compliance and that the large increases in shear modulus observed on fast neutron irradiation are due to covalent bonding between interstitial groups and the adjacent planes and not to the
pinning of dislocations.   Jenkins does not discuss in detail how
this hypothesis would be expected to influence current theories on
annealing kinetics and irradiation damage mechanisms in general, but
it is clear that arguments on the deformation process at the crystal
level will have to be resolved before a full treatment of fracture
can be formulated.

   The following series of experiments illustrate the importance
of basal plane shear and slip in the deformation behavior of graphite.
Soule and Nezbeda [24] studied basal plane shear in single crystal
graphite at room temperature and demonstrated the strong influence
of basal plane dislocation systems.   The stress-strain behavior
showed a marked basal plane slip activated at a low critical shear
stress $\sim 0.29 \times 10^6$ dyn/cm^2 (0.03 MN/m^2). $d\sigma/d\varepsilon$ was near zero at fracture, and typical values for the slip measured as the ratio of strain
at failure to the critical shear strain were in the range 38 to 10^5.
In comparison, the stress-strain behavior of compression-annealed
pyrolytic graphite was much more linear, with the grain boundaries
effectively restricting dislocation movement.   In an extreme example
of laminar flow, a single crystal exhibited a shear strain of 22
without fracture.   An average value of the shear strength was
4.8×10^6 dyn/cm^2 (0.48 MN/m^2).   Observations were made of delamination
voids of various types in the crystals, and also crack nucleation
from high stress concentrations at defects in the structure.   They
examined the effect of pinning dislocations by boron doping, which
produced a large increase in shear modulus, and concluded that the
intrinsic value of c_{44} for a dislocation free crystal was
$(0.45 \pm 0.06) \times 10^{11}$ dyn/cm^2, $(0.45 \pm 0.6) \times 10^4$ MN/m^2.

In experiments on compression-annealed pyrolytic graphite (a
highly ordered structure), Blakslee et al. [25] obtained a complete
set of elastic compliances from dynamic and static tests. The
static tests gave linear and reversible stress-strain curves in tension and compression, but nonlinear curves in the torsion tests for
the shear modulus c_{44}, with hysteresis and permanent set showing all
the characteristics of polycrystalline stress-strain curves and
suggesting that, at least initially, basal plane shear within the
crystallites is the source of this behavior in the polycrystalline
material. The shear strength of the basal planes was found to be in
the range 90 to 250 g/mm^2 (0.9 to 2.5 MN/m^2). Fast neutron irradiation of the same material at a low temperature was shown to increase
the shear modulus by an order of magnitude approaching a maximum
value $\sim$0.4 10^{11} dyn/cm^2, while having no significant effect on the
other elastic constants. Static torsion tests following the low
temperature irradiation produced considerably more linear stress-
strain relationships with only very small residual strains on unload.
These effects were attributed to the pinning of basal plane dislocations by the irradiation induced defects, as in Baker and Kelly's
experiments [20] on irradiated single crystals.

Apart from macroscopic strain measurements, there have been
several other studies of the events which precede complete failure in
polycrystalline graphite. These studies have taken the form of
acoustic emission studies, direct microscopic examination of regions
subjected to high stress levels, or determinations of the effect of
prestressing on properties.

B. Acoustic Emission

In their early studies of permanent set in cantilevered rods of carbons and graphites, Andrew et al. [26] reported that some carbons
emitted a "crackling" noise during initial stressing, but found that
the permanent sets could be removed by a thermal anneal. Gilchrist
and Wells [27] studied acoustic emission from two types of graphite
subjected to uniaxial compression. The British Reactor Grade A material gave positive indications of sound emission well before

complete failure. On stress cycling, the noises were only picked up
when the previous maximum stress was exceeded, and the authors concluded that they were caused by nonconservative processes, probably
internal cracking. Acoustic emission was not detected in the stronger
isotropic graphite until stresses were very near to the failure level.

C. Microscopic Examination

Many workers have examined fracture surfaces microscopically, but
the most valuable studies have been sequential microscopic examina-
tions of controlled crack growth.

Jenkins [28] studied controlled crack growth at room temperature
in a reactor grade petroleum coke graphite, bound and impregnated
with pitch, by bending thin strip specimens bonded to a brass strip
and observing under a microscope the sequence of events in the plane
of maximum stress subjected to increasing uniform strain. This
graphite comprised comparatively large and well-oriented needle-like
grist particles in a binder matrix of more random structure. Fracture was observed to proceed preferentially along striations within
the highly oriented structure of the grist particles, particularly
under shear at angles to the direction of maximum strain. Cracks
traveled between pores, and isolated cracks appeared before major
fracture. He concluded that the material deforms mainly by interlamellar slip and fracture within the grist particles inhibited by
the less well-oriented binder and that crack propagation is considerably more difficult than crack initiation.

Slagle [29] used a controlled brittle ring test on specimens of
a Texas coke graphite SGBF to observe the sequential propagation of
cracks under increasing load. In this test, the ring is loaded in
compression across its diameter, and high stress gradients are established. Separate regions with cracks propagating down tensile and
compressive stress gradients were examined microscopically. Cracks
appeared to initiate preferentially at the boundaries of existing
pores, again following an interlayer path, and in the tensile region
were first observed at about $0.5 \sigma_t$ where σ_t is the stress at which
complete failure occurs. There was evidence for a build-up in crack

density as original cracks stabilized at pores and new ones formed.
After complete failure, some of the cracks near the major failure
path closed up, but others stayed open, suggesting relief of initial
internal stress.

Using a different technique, Taylor et al. [30] and Knibbs [31]
also subjected graphite samples to a controlled tensile strain, induced by differential thermal expansion with a heated metal holder,
and progressively examined the region around propagating cracks.
Taylor et al. examined isotropic reactor grade graphites based on
petroleum coke fillers with coal tar pitch binder, with a maximum
particle size of ∿1 mm. Cracks initiated at pores, and the major
fracture path linked the larger pores. Crack branching was observed.
Crack extension sometimes occurred by secondary cracks initiating
at pores ahead of the main crack and traveling back to meet it.
Spherical coke particles tended to deflect the path of cracks around
them. Knibbs studied three widely different graphite types differing in grain size, namely, a coarse-grained nuclear grade material,
Morgan's EY9, and a fine-grained POCO graphite AXF. He concluded
that in the coarse- and medium-grained materials, containing filler
grains of well-aligned crystallites, the weakness was associated with
these grains with cracks cleaving their ordered structure or running
along particle binder boundaries. Secondary cracking was particularly
prevalent in regions of high pore density, and pores tending to
stabilize crack growth. Anisotropy in strength was attributed to the
more tortuous crack path in the high strength direction, with the
cracks deflected around aligned filler particles. It was not easy
to identify the weak link in the more homogeneous structure of the
fine-grained material.

Oku and Eto [32] show microphotographs of the successive growth
of cracks in three nuclear grade graphites (IM-2, 7477 PT, and H-327)
subjected to increasing compressive stress. Crack growth becomes
noticeable above a stress level of about $0.6\sigma_c$, where σ_c is the compressive fracture stress.

Jouquet [2] has also described microscopic observations of the
generation and propagation of cracks in different graphite structures.

In anisotropic graphites containing highly oriented grains, fracture
paths cleave the grains or pass around their boundary, depending on
the grain orientation with respect to the applied stress, and in
graphites based on Gilsonite coke, which has a spherical structure,
cracks tend to pass around the grains. The strain at which the microcracking is first detected depends on the material studied. In
needle coke graphites of large grain size, the level was about 0.1%,
but was some three times greater in a very fine-grained material
for which the pores appeared to play a prominent part in determining
the fracture path.

Meyer and Buch [33] have described an experimental program on
different types of graphite designed to observe crack initiation and
growth under stress in situ with an electron microscope. Their results are similar to the observations reported previously from optical
microscopy and show that propagation of cracks is coincident with
highly ordered directions within the filler particles. The short
range cracks join to form a large crack as the strain increases.
While the orientation of these cracks in particular localized regions
can deviate considerably from the tensile plane, by as much as 50°,
the average orientation over the length of the total crack path only
deviates by a small amount, about ±5°. The role of pores is also
important in acting as centers for crack termination and emission,
particularly in the finer-grained materials.

In their studies of high temperature graphite creep, Green et al.
[34] developed a method of sequential optical and electron micropho-
tography of the same area on graphite specimens subjected to tensile
stress at high temperatures. Their observations on ZTA graphite before and after increasing creep strains at 2500°C show similar conclusions, with creep strain resulting from cracks which propagate
mainly between pores in a direction mainly parallel to the average
basal plane orientation. Their electron micrographs failed to detect basal or nonbasal slip bands and grain boundary sliding. The
major part of the elongation was due to crack formation and growth,
and cracks identified as due to basal plane cleavage appeared to
contribute only a small fraction of the overall strain.

Their work [35] in the temperature range 2200 to 2500°C on both POCO graphite grade HDP-1 and a typical commercial graphite showed that in the latter, creep failure occurred by the opening of cracks in regions between the larger pores, the crack system eventually linking and causing failure. Smaller pre-existing pores were unchanged in shape, and no crack formation between them was evident. In the POCO material, which was free of large pores, the extension was associated with the opening and growth of small pores. Lack of large pores and hence higher density of small pores resulted in a higher fracture stress and higher extension to failure in the POCO graphite. Green et al. [36] have also demonstrated the build-up in microcrack concentration with tensile strain at 2500°C in the region around an artificial defect machined in the graphite sample.

Smith [8] has given an excellent qualitative summary of the observations on microfracture discussed in this section, describing the fracture process as a build-up of nonpropagating microcracks preferentially in regions of high stress concentration adjacent to the large "disparate" voids, until a microcrack density is reached where they join to form a discontinuous macrocrack of critical size which then propagates.

D. Effects of Prestress on Young's Modulus and Strength

Associated with these visual observations of microcrack formation prior to ultimate failure, there have been several investigations of the effect of prestressing on Young's modulus and a few studies of the effect on strength. These studies have invariably shown a decrease in modulus with prestressing, all leading to discussions on the relative contributions from increase in dislocation density (or unpinning effects), and the formation and propagation of cracks.

Losty and Orchard [11] showed that preloading in compression of a reactor graphite to 21 MN/m^2 (probably about three-quarters of the fracture stress) reduced the elastic modulus but did not significantly affect the average tensile strength on subsequent tests.

Hall [37] showed that the modulus decrease caused by compressive prestressing of a similar graphite would recover at least partially after thermal annealing or exposure to fast neutron irradiation, or even spontaneously with time at room temperature. After stressing to 21 MN/m^2, however, a significant fraction of the 35% reduction was not recovered by thermal annealing. He concluded that at low prestress levels the fall in modulus was due entirely to an increase in dislocation density but microcracking contributed at higher stresses.

Hart [38] studied changes induced by prestressing ATJ and AXF graphites in both tension and compression to 90% of the failure stress. Density and thermal expansion were reduced by tension and increased by compression, but dynamic Young's modulus was reduced in both stress modes with the major effect in compression. The modulus changes caused by prestressing were removed completely in AXF and partially in ATJ graphite by thermal annealing at 1300°C, so that if the effects are associated with crack opening, closure and healing must occur on annealing. He attributed the modulus changes to an increased dislocation density and/or the introduction of additional cracks which in compression do not affect thermal expansion, only modulus.

Oku and Eto [32] accompanied their direct observations of crack growth in compression with both modulus and strength measurements on prestressed samples. For three basic types of graphite, they found that dynamic modulus progressively decreased with increasing prestress level. Changes were small at low prestress levels, but the modulus decreased more rapidly after compressive prestress levels >0.2 to 0.3 σ_c, where σ_c is the compressive strength. Subsequent changes near the fracture stress were greater in the anisotropic needle coke graphites (H-327, SMG) than in the isotropic materials (IM-2, IE 1-24, 7477PT). The average compressive strengths of the graphites were unaffected by prestress level or strain rate (in the range 10^{-4} to 10^{-2}/sec). However in some tests, the tensile strength was reduced by compressive prestress, although the effect differed for each material and was also dependent on strain rate. For example, in the

Gilsocarbon graphite IM-2, the tensile strength was substantially
unchanged at a strain rate of 7.9 10^{-2}/sec over the full prestress
range but showed a small but significant decrease at prestress levels
>0.6 σ_c for a lower strain rate of 7.9 10^{-5}/sec. At this lower rate,
the tensile strength of the anisotropic needle coke graphite H-327
showed no change up to $0.5\sigma_c$, abruptly decreased to about one-third
of the value, and then changed little at higher prestress levels.
Oku and Eto could not conclusively show whether an increase in dislocation density or the formation of cracks gave the major contribution to modulus decrease with increasing prestress, but in their
analysis of the data, they favored the crack mechanism, particularly
at high stress levels.

In support of this mechanism, it should be noted that for the
two materials showing the greater changes, the decreases in strength
were greater than the corresponding decrease in modulus. Consideration of the Griffith [39] equation relating modulus and strength
with flaw size predicts $\sigma \propto (E)^{\frac{1}{2}}$ for no change in flaw size c
(i.e., dislocation effect only), and hence predicts smaller changes
in strength than in modulus, but gives $\sigma \propto (E/c)^{\frac{1}{2}}$ with an increasing
crack c and could then explain the larger changes observed in σ than
in E.

The more recent work [40] of these authors on electrical resistivity changes during compression support the general observation
that at stresses >$0.5\sigma_c$ the formation and growth of optically resolv-
able cracks occur. At low stresses, the resistivity of with-grain
specimens increased in contrast to the decrease observed for against-
grain specimens, due to the different deformation characteristics
of the preferentially aligned cracks and pores in the two directions.

Both Jenkins [28] and Matthews [41] also showed a progressive
decrease in dynamic Young's modulus with increasing value of prestress.
Jenkins examined the effect of compressive prestress on a reactor
grade material based on petroleum coke filler and pitch binder, and
Matthews studied the effects of prebend stress on the fine-grained
POCO AXF-5Q graphite. Matthews further demonstrated that prestressing AXF-5Q graphite in 4-point bend to a level near the mean failure

stress for the batch produced significant damage in the surviving
specimens, in that subsequent bend tests produced failure in 20% of
the specimens at a level below the initial proof stress.

In bend tests on CFW graphite, Dally and Hjelm [42] concluded
that proof testing to 95% of the median strength level did produce
some damage in that about 4% of the survivors broke slightly below
the proof level. However, working with the AXF-5Q material, Jortner
[43] found that in both tensile and compressive tests proof testing
to 90% of the average ultimate strength followed by annealing at
2000°C in helium did not cause any detectable damage in the survivors
of the proof test, thereby offering a method of improving the strength
distribution of service components. The annealing completely removed
the permanent sets established in the proof test.

Closely allied to these effects are the fatigue characteristics
of graphite. Dynamic fatigue, which is believed to be caused by a
crack growth mechanism, is discussed in detail in a later section,
but graphite will also exhibit a time dependent failure under constant load. For example, Wilkins [44] has shown that RC4 graphite
exhibits this effect of "static fatigue" at stresses some 5% below
those required to cause instantaneous fracture. Recent work by
Hodkinson and Nadeau [45], using a fracture mechanics approach on
POCO graphites AXZ, AXF-5Q, and pyrolytic graphite, has shown that
little slow crack growth occurs at room temperature except at very
high fractions of the critical stress intensity.

E. *Micromechanisms of Failure*

The work described in these early sections indicates that the experimental observations of events occurring when graphite is stressed
are a consequence of dislocation activity at low stresses and a progressively increasing contribution from microcrack generation as the
stress is increased. The difficulty of separately identifying these
phenomena and assessing their relative contributions are their close
association and the statistical overlap in their observed effects on
the material properties. The observed behavior suggests that failure

is essentially brittle in character, but crack initiation may be
preceded by plastic deformation.

In perfectly elastic materials, crack initiation at inherent
flaws and subsequent propagation is governed by a Griffith [39] crack
mechanism in which the energy balance criterion interrelates the
fracture strength, modulus, defect size, and surface energy absorbed
per unit area exposed by crack propagation. The surface energy for
graphite crystals is about 0.15 J/m^2 for basal planes in air [46,47]
and about 6 J/m^2 [46] for crystal surfaces perpendicular to the basal
planes. Figure 1 attempts to illustrate the variation in the rate
of energy demand for a crack passing through the complex polycrystalline structure containing pores, with the crystals having these widely
anisotropic surface energies. The exact sequence of events will be
influenced by the loading conditions, i.e., whether constant stress
or constant strain applies. While cracks travel preferentially
parallel to the basal planes in directions approximately normal to
the applied stress, a proportion of the fracture path creates nonbasal surfaces with a high rate of energy demand, and crack branching
also absorbs energy.

In materials for which plastic deformation by slip in the high
stress regions at the tips of pre-existing flaws occurs prior to
ultimate failure, crack initiation is generally a consequence of
stress concentrations arising from dislocation pile-up at obstacles
or from a limitation in the number of available slip systems. The
von Mises [48] criterion that five independent slip systems are
required for a crystal to undergo a general homogeneous strain by
slip is discussed by Groves and Kelly [49], who point out that graphite
deforms readily by slip along its basal planes defined by only two
independent systems. This limitation causes accommodation problems in
a dense polycrystalline aggregate arising from the inability to match
strain components at grain boundaries because of orientation differences. The resulting stress concentrations can cause rupture controlled by the surface energy unless either slip on a new system is
initiated by the high stress or diffusion processes operate to relieve

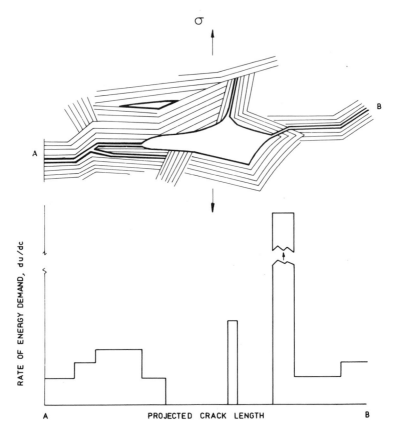

FIG. 1. Rate of energy demand for a crack moving through polycrystalline graphite from A to B.

the stresses before cracks start to open up. Alternatively, the
presence of porosity suitably disposed for grains to deform into
may partially relieve these problems of accommodating plastic deformation and delay the onset of fracture. Hence microfracture by this
type of mechanism would be expected to be influenced by those factors which affect the accommodation of the deformation of elemental
components of the structure. At the gross structural level in a
polycrystalline material, such factors are the density of the material, or more specifically that contribution which determines the
porosity available to absorb grain deformation, as well as its

grain size, which determines the surface boundary to volume ratio of
the deforming element.

While there appears to be common ground in the qualitative descriptions of the events which precede ultimate failure in polycrystalline graphites, there is no universally agreed quantitative description of the detailed mechanisms involved. Attempts to relate
observed strengths with fundamental characteristics of the material
have been made at various structural levels.

Reynolds [22] argued for a crack mechanism of failure. For
well-graphitized materials with elastic deformation governed by
basal shear, he considered the appropriate modulus for use in the
Griffith relationship to be the effective crystal shear modulus c_{44}
reduced by the presence of basal plane dislocations. With the basal
plane surface energy he obtained a critical defect size of $\sim 10^{-3}$ cm,
in agreement with the magnitude of the basally oriented microcracks
observed in electron microscope studies [7] and attributed to
anisotropic contraction on cooling from the graphitizing temperature.

Mason [50] also supported a crack theory of failure but at a
macroscopic level. He pointed out that the macroscopic properties
of modulus and strength were consistent with a defect size of $\sim 10^{-1}$ cm,
comparable with the grain size in the material he considered: The
highly oriented structure of the grain was considered to be the weak
link. Taylor et al. [30] identified the maximum pore size (comparable with the grain size in the materials investigated) as the characteristic defect in a Griffith type of relationship with macroscopic
properties. At the other end of the scale, Virgil'ev et al. [51]
have correlated the compressive strength σ_c of a heat treatment
series of petroleum coke materials with the crystallite size L_a.
For the less well-graphitized materials, the relationship was found
to be described by the Petch [52] formula:

$$\sigma_c = \sigma_o + kL_a^{-\frac{1}{2}}$$

where σ_o is the frictional resistance to dislocation motion and k is
a constant.

Jenkins [1] has stated that in materials exhibiting a high
degree of interlamellar shear, a dislocation pile-up mechanism probably operates, but in materials with a high degree of elastic restraint, the Griffith crack mechanism probably applies. It seems
likely that the initial generation of nonpropagating microcracks is
governed by surface energy, following stress concentration at the
tips of preexisting flaws, either directly in an elastic situation
or by processes associated with plastic deformation of the type discussed previously. Interlinking of these microcracks cause them to
attain a critical length with ultimate failure of the material determined by a Griffith type of relationship between the macroscopic
properties involving an effective work of fracture.

The Griffith model has in fact been the main yardstick against
which the macroscopic fracture properties have been compared, and the
application of linear elastic fracture mechanics to graphite has
received considerable study in recent years. As is discussed later,
macroscopic measurements of the effective work of fracture are
several orders of magnitude greater than the single crystal surface
energies (10^2 J/m^2 compared to 10^{-1} J/m^2) attributed to the absorption of energy by multiple crack formation and the tortuous crack
path, and probably also a contribution from fracture along nonbasal
planes, the proportion of which will vary with the type of material.

The ideal would be a quantitative understanding of the microcracking which precedes ultimate failure and its relationship with
the conditions for the latter via measurable features of the microstructure. A significant step in this direction lies in the work
described by Meyer and Buch [33] to back up their theoretical model
[53] of failure by coincidental alignment of defects within the
graphite structure following the pattern of experimentally observed
crack formation described earlier in this section. The basic defects
are considered to be cleavage planes along basal directions in the
filler particles, so that coincidental alignment determines the size
of the effective crack and ultimately the failure stress. The
distribution and size of the pores are taken into account in addition
to the size and orientation of the filler grains. The development

of this model is aimed at furthering the understanding of the fracture process and defining the important microstructural features, so
that a description is obtained consistent with experimental observations and data, and so that graphites may be tailored to suit different applications.

Anderson and Salkovitz [54] have proposed a semi-empirical fracture criterion based on Griffith theory, which takes account of
porosity and grain size variations in different graphites.

Any overall model must describe the macroscopic behavior of the
material of importance to the engineer and indicate the way in which
laboratory test results should be applied in real service situations.
The following sections consider the macroscopic failure of graphite
under a variety of situations in an attempt to give a complete picture of failure under different stress modes and the influence of
different environments, thus describing the behavior which must be
accounted for by any model used in engineering design.

IV.   FRACTURE UNDER UNIFORM STRESS

This section discusses the limiting states of stress defining fracture under multiaxial stress fields which are substantially uniform.
It is common practice to characterize the uniaxial strength of
graphite under uniform tension or compression, but the strength under
combined stress systems has also been determined. These multiaxial
stress tests have generally taken the form of either solid cylinders
subjected to axial loading with variable radial compression, or
thin-walled hollow cylinders axially loaded with internally or externally applied pressure to generate circumferential tensile or
compressive stresses, respectively. In such tests on hollow cylinders, the geometry is chosen so that both the circumferential stress
gradients and the radial stresses through the wall are small enough
to be neglected in considering the fracture behavior. Fracture under
nonuniform conditions is discussed in a later section.

First, however, the behavior of an ideal brittle material is
considered.

A.   The Griffith Model

The Griffith model [39] of brittle fracture is based on the hypothesis that failure is due to stress concentrations caused by the presence of inherent crack-like defects. Graphite has an abundance of such potential defects at macroscopic and microscopic levels, and before discussing the graphite data further, it is useful to summarize the relevant conclusions resulting from the development of this model in order to give a basis for comparison with graphite behavior.

Under a plane tensile stress the Griffith-Orowan theory for the failure of a body containing sharp cracks of length 2c gives the critical stress at fracture

$$\sigma \sim \left(\frac{2E\gamma}{\pi c}\right)^{\frac{1}{2}} \tag{1}$$

where

> γ = effective surface energy per unit area
> E = elastic modulus

Following his original paper, Griffith [55] gave the solution for a biaxial stress state with cracks of uniform size randomly oriented in the principal stress plane. With the hypothesis that fracture results when the maximum stress on the periphery of the crack reaches a critical value as defined in uniaxial tension, this approach leads to the following criteria for failure under biaxial stresses, discussed for example by Murrell [56]. For $3\sigma_1 + \sigma_2 > 0$

$$\sigma_1 = \sigma_t$$

and for $\sigma_1 - \sigma_2 > 0, \quad 3\sigma_1 + \sigma_2 < 0$

$$(\sigma_1 - \sigma_2)^2 + 8\sigma_t(\sigma_1 + \sigma_2) = 0$$

$$\left.\vphantom{\begin{array}{c}1\\2\\3\end{array}}\right\} \tag{2}$$

where σ_1 and σ_2 are the major and minor principal stresses, respectively, and σ_t is the uniaxial tensile strength. The theory predicts that the uniaxial compressive strength of a brittle material containing sharp cracks is eight times its uniaxial tensile strength. This ratio is approached in certain ceramics which show linear elastic stress-strain behavior to failure.

McClintock and Walsh [57] modified the Griffith theory to allow
for crack closure in compression and showed that compressive/tensile
strength ratios greater than eight may be obtained due to frictional
forces operating on the crack face. This result is particularly
relevant to the failure of rocks.

Babel and Sines [58] also modified the Griffith theory to allow
for elliptical defects of different shape and predicted that the
compressive/tensile strength ratios may vary between limits of three
to eight depending on the sharpness of the relevant cracks, with
spherical voids giving a value of three and infinitely sharp cracks
giving the value of eight.

For a solid containing sharp Griffith cracks, Eqs. (2) define
the fracture envelope under biaxial stress. The criterion in the
tensile-tensile (T-T) quadrant and the upper part of the compressive-
tensile (C-T) quadrant is equivalent to a maximum principal stress
theory. In the lower part of the C-T quadrant, and extending into
the compressive-compressive (C-C) quadrant, the envelope is parabolic
and has been shown by Murrell [56] to be equivalent to the Mohr
envelope in this region. The model predicts an increasing strength
under an increasing normal compressive stress for a material containing cracks oriented only in the biaxial stress plane. It is probable,
however, that in this C-C quadrant the strength of a real material
with flaws randomly oriented in three dimensions, i.e., removing
the planar restriction on the above model, is limited by the uniaxial
compressive strength, since failure would occur at flaws suitably
oriented to be unaffected by one of the two principal stresses. This
argument leads to a biaxial failure envelope for a Griffith solid as
indicated in Fig. 2.

Real materials also contain distributions of flaw sizes requiring
a statistical treatment. One particular approach to the statistics
of extreme variables is that due to Weibull [59] and is discussed in
detail later.

Failure in tension of a Griffith solid can occur by the propagation of a single crack according to Eq. (1), with the resultant
fracture face normal to the applied stress. The development of

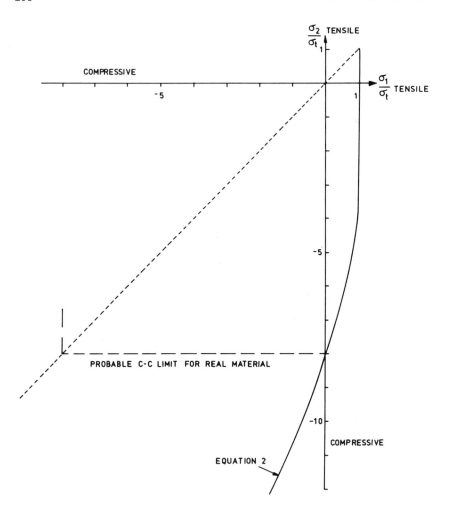

FIG. 2. Biaxial failure envelope for a Griffith solid.

cracks in compression is more complex and is discussed, for example,
by Obert [60]. As observed in glass by Brace and Bombolakis [61],
the cracks propagate from near the tip of a critically oriented
flaw angled to the principal stress, thereby tending to run out of
critical orientation. The compressive fracture plane of truly
brittle materials is frequently observed to be parallel to the
applied stress.

B. *Uniaxial Strength*

The tensile strengths of commercial polycrystalline graphites at
room temperature are typically in the range 10 to 30 MN/m^2, although
some special high strength materials have strengths at least a factor
of two higher than this value. Fracture surfaces are normal to the
applied stress. Typical values of the total tensile strain to failure
of the moderate strength materials are in the range 0.1 to 0.3%, of
which a substantial fraction remains as a permanent deformation at
room temperature, but again some high strength graphites have strains
to failure well above this range.

The Griffith equation (1) formally relates the important parameters but as is seen later, measured effective surface energies are
orders of magnitude greater than the theoretical value due to the
absorption of energy in the fracture processes already discussed,
i.e., plastic deformation/multiple cracking, crack branching, and
the tortuous crack path.

Inserting into Eq. (1) typical values of the macroscopic prop-
erties $\sigma_t \sim 20$ MN/m^2, $\gamma_v \sim 5 \times 10^4$ erg/cm^2, $E \sim 10^4$ MN/m^2, the length
of the inherent defect size is about 1 mm, a characteristic size
which is of similar magnitude to the grain size and/or pore size in
real materials with the foregoing property values. The relationship
between strength and grain size and the concept of these relatively
large inherent flaws is discussed later.

Compressive strengths are typically higher than the tensile
strength by factors of three to four, considerably lower than the
value of eight predicted by the Griffith model. At first sight,
this observation suggests that the defects may be regarded as blunt
and may possibly be modeled on the Babel and Sines modification to
the Griffith theory referred to earlier. Taylor et al. [30] showed
that the ratio of compressive to tensile strength increases with
irradiation damage (as stress-strain curves become more linear and
"plastic" deformation is reduced) from the unirradiated value of
three to four toward the value of eight predicted by the Griffith
theory. If the tip radius can be regarded as a measure of the

extent of the damaged zone at the crack tip, this result could be
interpreted as the effective "sharpening" of cracks by irradiation
caused by a decrease in plastic deformation (or associated microcrack formation) in the material near the crack tip. This offers
a qualitative argument, but the application of the Babel and Sines
model to biaxial strength data is discussed in the next section,
where it is shown to give a poor description of the failure surface.

Graphite compressive tests are very often reported to give a
"shear" type failure with the fracture plane angled to the applied
stress direction (e.g., Gillin [62] reports an angle of 45°, and
Taylor et al. observed about 35° [30], although in tests on some materials longitudinal splitting has been observed [13]). The fracture
process has already been described as resulting from the linking of
coincidental microcracks which form in high stress regions and stabilize, rather than the propagation of a single crack. Failure under
compression can be visualized as the linking of cracks along a plane
of maximum net shear stress, although observers of longitudinal
splitting have argued for a lateral tensile strain failure.

Taylor et al. [30] observed the shear strength of two graphite
types to be about twice the tensile strength (i.e., about half the
compressive strength). Other workers have quoted values of about
half the tensile strength, but this appears to imply that the
graphite will not withstand uniaxial stresses greater than the tensile
strength in either tension or compression, which is clearly not observed.

It is interesting to note that some glass-like carbons [63] have
tensile strength values as high as 200 MN/m^2 and corresponding compressive strengths of about 1400 MN/m^2, a ratio of seven which is
closer to the Griffith value of eight, indicating that this material
may well have the characteristics approaching that of the ideal
Griffith solid.

C. Biaxial Strength

Table 2 summarizes several sources of graphite strength data obtained
under substantially biaxial stress fields using tubular specimens
subjected to combined internal pressure and axial loads. In most

TABLE 2

Sources of Graphite Biaxial Strength Data

| Reference | Graphite | Approximate sizes (in.) | | Quadrant investigated |
		Inner diameter	Wall thickness	
64	Graph-I-Tite, A and G	1	0.06	T-T, C-T
65	Graph-I-Tite, G	1	0.06	T-T, C-T
66	Graphi-I-Tite, G	1.5	0.125	T-T, C-T
67	ATJ-S	1	0.05	T-T, C-T, C-C
68	AXF-5Q	1	0.05	T-T, C-T
69	ATJ	0.9	0.05	T-T, C-T, C-C
70	ATJ-S	2	0.25	T-T, C-T, C-C

experiments, the specimens were sleeved to prevent penetration of
the pressurizing fluid, and the relatively small radial stresses are
justifiably ignored. Some of these experiments also give stress-
strain data.

The data from all these experiments show a similar pattern, although there are some differences due to variations in degree of isot-
ropy and, possibly, experimental problems in achieving the required
stress pattern. However, discussion of the results allows the typical
characteristics for an isotropic graphite to be described and compared
with the theoretical model developed earlier.

In the T-T quadrant, it can be argued that some of the data in
the literature fit the Griffith criterion in Fig. 2, although there is
experimental difficulty in obtaining data near the equibiaxial position to define the degree of sharpness in the corner of the square.
However, an overall general view is that the mean biaxial strength
tends to be lower than the uniaxial tensile strength near the point
of equal stresses. Jortner shows definitive data in this quadrant
for both ATJ-S graphite [67] and also for the isotropic graphite
AXF-5Q [68] for which the equibiaxial condition is defined by
$\sigma_{biaxial} \approx 0.85\sigma_{uniaxial}$. The physical interpretation in terms of
a flaw mechanism of failure is that under a biaxial stress there is
a greater chance of the larger flaws being critically oriented to
a principal stress.

Jortner points out in both of his papers that fracture criteria must take account of the statistical chances of survival, and draws attention to the fact that his lower limit data sensibly obey a maximum stress criteria of failure. This observation is consistent with the interpretation that the largest flaw determines the same minimum strength whether oriented normal to the uniaxial stress or to one of the two components of the biaxial stress. To apply this to design problems, a clear definition of the lower limit must be made.

In the C-T quadrant, the tensile fracture stress is largely un-affected by applied compressive stresses at least equal in magnitude to the tensile strength σ_t (shear condition), but then decreases toward zero as the compressive stress approaches the uniaxial com-pressive strength σ_c. Most of the authors in Table 2 give a fairly well-defined envelope in this quadrant. Figure 3 shows normalized data in this quadrant for a near-isotropic graphite, grade IE 1-24, which shows the same characteristics as the data reported in the literature. Comparison with the Griffith theory shows that the main difference lies in the ratio of the uniaxial strengths as discussed earlier.

In the T-T and C-T quadrants, Jortner [68] observed that for AFX-5Q graphite, as expected from Poisson's ratio effects, tensile hoop strains at failure were reduced by axial tension and increased by axial compression. The effect was not as marked in his studies on graphite ATJ-S [67]. The radial strain is not negligible, however, and the two-dimensional plot does not give a comprehensive picture. Jortner also showed that the effect of temperature up to 4000°F was to raise the biaxial strength by approximately the same ratio as for the uniaxial tensile strength.

In the C-C quadrant, the biaxial data form a closed loop al-though there is considerable scatter, and some experimentalists ex-press reservations about their data due to difficulties in avoiding premature failure by buckling. However, Jortner's [67] data for

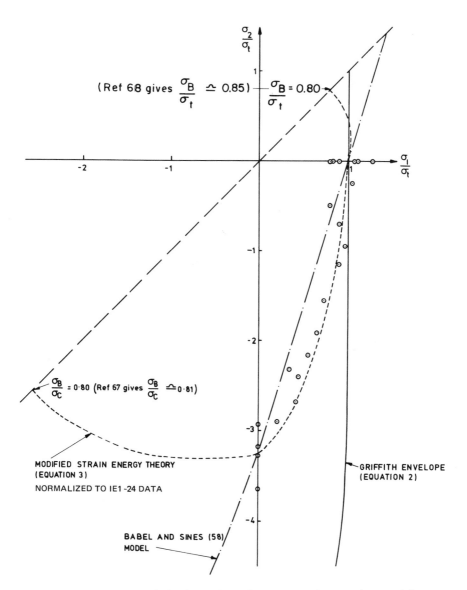

FIG. 3. Biaxial failure data for a near isotropic graphite
(IE1-24) (normalized to uniaxial tensile strength $\bar{\sigma}_t$ = 13 MN/m^2),
compared with empirical models and other data.

ATJ-S graphite at the equibiaxial point is free from this reserva-
tion and shows that the equibiaxial strength $\sigma_B \approx 0.81\sigma_c$, where
σ_c is the uniaxial compressive strength.

Thus the biaxial strength data for unirradiated graphite do not
conform to the Griffith model for a perfectly brittle material
(Fig. 2) largely in that the ratio of uniaxial compressive to tensile strength σ_c/σ_t, is considerably lower than the predicted value
of eight, which modifies the envelope, and that there is some reduction in strength under near-equal biaxial stresses. Figure 3 shows
that the Babel and Sines modification to the Griffith mode, which
ascribes lower ratios of σ_c/σ_t to the shape of the inherent defects,
does not give a good fit to data in the C-T quadrant for low values
of this ratio, nor does it describe the experimental observations in
the T-T quadrant. Taylor et al. [30] report that under fast neutron
irradiation at 150°C, the ratio σ_c/σ_t increases from the low unirradiated value toward the value of eight predicted by the Griffith
theory. Under these irradiation conditions, the stress-strain curves
become more linear as plastic deformation is reduced, and the material behavior tends toward the brittle Griffith type solid.

Statistical arguments can be offered to explain mean strength
reductions under biaxial tension. Price and Cobb [71] have shown
that the Weibull theory predicts the observed fracture envelope in
the T-T quadrant, and they evaluated the envelopes corresponding to
Weibull modulus m values in the range 2 to 20. A value of m = ∞ for
a perfectly homogeneous body corresponds to the Griffith (or maximum
stress) criterion of failure in this quadrant. Their calculations
show that the strength under equibiaxial stresses ($\sigma_B \approx 0.85\sigma_t$) is
achieved with an m value of approximately 12, but less homogeneous
materials with lower m values would show more substantial reductions
in biaxial strength. Figure 4 shows the ratio of equibiaxial strength
to uniaxial tensile strength as a function of the Weibull modulus m
from Price and Cobb's calculations. However, the Weibull theory
does not apply successfully in the C-T quadrant since it predicts
[58] an increasing tensile fracture stress with increasing compressive stress. It is interesting to note that this behavior reflects

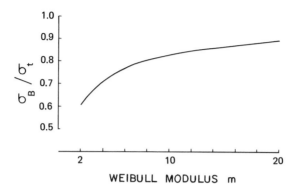

FIG. 4.   Ratio of biaxial to uniaxial tensile strength as a function of Weibull modulus (calculated from data in Ref. 71).

that of the strain to failure envelope described for AXF-5Q graphite by Jortner [68].

In view of the inability of physically based theories to describe the biaxial stress failure envelope successfully, various empirical theories have been proposed for engineering design use. Most of these theories are in the form of quadratic equations expressed in terms of the principal stresses with the constants defined by the uniaxial strengths. Allowance is made for in-plane anisotropy in particular materials by treating each of the four quadrants separately. An equation of this type which gives a reasonably good description of much of the experimental data in the T-T and C-T quadrants, and possibly gives a good representation in the C-C quadrant also, is the modified maximum strain-energy theory described by Ely [65].

$$\left(\frac{\sigma_1}{\sigma_t}\right)^2 - 2\nu\,\frac{\sigma_1\sigma_2}{\sigma_t\sigma_c} + \left(\frac{\sigma_2}{\sigma_c}\right)^2 = 1 \tag{3}$$

where

$\sigma_1\sigma_2$ = principal stresses
$\sigma_t\sigma_c$ = corresponding uniaxial strengths
ν = Poisson's ratio

Figure 3 shows the failure envelope generated by Eq. (3) using a typical value of Poisson's ratio of 0.2. The curve describes the

experimental data in the C-T quadrant better than either the Griffith
model or the Babel and Sines modification. It also gives the equibiaxial points in the T-T and C-C quadrant in approximate agreement
with the experimental data of Jortner [67, 68]. It should be noted
that some of the data in the literature give biaxial strengths considerably lower than suggested by the curve in Fig. 3, but their statistical significance is not clear. The shape of the curve generated
by Eq. (3) is not very sensitive to the value of Poisson's ratio used
within values (0.1 to 0.3) normally attributed to different graphites.
The ratio of equibiaxial to uniaxial strength varies from about 0.75
to 0.85 for Poisson's ratio in the range 0.1 to 0.3.

This type of theory, while giving a phenomenological account of
the mean strength envelope, does not account for the statistical
variations in failure stress which are of more concern to designers
than mean strength values. A new approach has recently been made
by Batdorf and Crose [72] to develop a physically based statistical
theory of fracture for isotropic brittle materials which assumes
fracture only depends on the macroscopic stress normal to a crack
plane. The theory is applicable to arbitrary stress states which are
not predominantly compressive and is suitable for finite element
techniques since the probabilities of failure of each element can be
compounded to predict that of the whole body. The theory was tested
by the authors against the biaxial strength data for AXF-5Q graphite
and showed good agreement. Hopefully it can be demonstrated that
the theory will predict the observed behavior under other stress
states including nonuniform stress situations, which are discussed
later.

D. *Triaxial Strength*

Strength measurements have also been made under triaxial stress conditions. Jortner [67] reported that the tensile failure stress of
ATJ-S graphite is decreased by equal compressive stresses applied in
the other two directions.

Taylor et al. [30] demonstrated that radial confining pressures
increased the axial compressive strength of sleeved cylindrical

specimens for three types of near isotropic graphite and PGA material.
Over the range of stress examined, the data fit the linear relation
given by the empirical Coulomb-Navier theory [73] extremely well.
The basis of this theory, which is in fact a biaxial theory, is that
the shear stress Γ across the fracture plane is opposed by a frictional stress $\mu\sigma$, where σ is the stress normal to this plane and
μ is a coefficient of friction. The shear strength S is thus increased and is given by

$$S = \Gamma + \mu\sigma$$

which leads to the expression

$$\sigma_1[\mu + (1 + \mu^2)^{\frac{1}{2}}] + \sigma_2[\mu - (1 + \mu^2)^{\frac{1}{2}}] = 2S_o = 2(\sigma_1 - \sigma_2) \quad (4)$$

where

σ_1 = hydrostatic pressure
σ_2 = axial stress (tensile stresses positive)
μ = 1/tan 2θ defining the inclination of the fracture plane

The modification to the Griffith theory by McClintock and Walsh [57],
mentioned earlier, gives essentially the same equation. Taylor et al.
found good agreement between measured and calculated fracture angles
which were similar for all four graphites (about 37°), independent
of hydrostatic pressure, giving a mean μ value of about 0.3. Fast
neutron irradiation increased this coefficient and decreased the
calculated fracture angle.

To illustrate the extent of the experimental observations under
triaxial stresses, Fig. 5 shows data obtained on sleeved specimens
of a near isotropic graphite (grade IE 1-24) for the compressive
fracture strength under radial confining pressures both before and
after irradiation at 150°C to a fast neutron dose of 3 x 10^{20} n/cm^2
(DNE). The data, which are normalized to the unirradiated tensile
strength, are plotted in the symmetry plane $\sigma_1 = \sigma_2$. In the tensile-
compressive region of this plane, Jortner's data would indicate an
envelope of the type shown by the dotted line. The stress locus for
equitriaxial stresses also lies in this plane, and the results of
experiments under such conditions are discussed later.

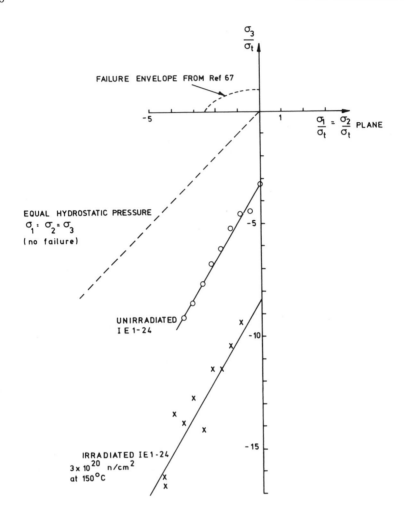

FIG. 5. Triaxial failure data for graphite

The behavior of graphite under high hydrostatic pressure has been reported by Kmetko et al. [74]. Sleeved specimens, with the pressurizing fluid excluded from the pores, showed a very large reduction in porosity (similar to the squeezing of a sponge) and remarkable recovery of strain, with only a small permanent set, on removal of the stress. For example, their clad specimen of AUC graphite showed a length decrease >10% for hydrostatic pressures up

to 15 kbar. The length changes were nonlinear with pressure, and
there was nearly complete recovery on removal of the stress. Similar
observations were made by Paterson and Edmond [75] in their high
pressure studies of the stress-strain behavior of EY9 Electrographite
in which axial stress was independent of the radial confining pressure. When the pressurizing fluid was allowed to enter the pores,
the measured strains were considerably smaller, and in Kmetko's
work the behavior appeared to be substantially elastic. At lower
stresses, Jortner [67] has also studied the behavior of the ATJ-S
graphite under hydrostatic pressure.

In a nonporous body, shear stresses are not produced by hydrostatic pressurization, and only elastic behavior is exhibited. In
a sleeved porous body, with the pressurizing fluid excluded from
the pores, localized shear stresses are established, and plastic
deformation due to dislocation glide can occur. Pore collapse by
localized fracturing would also be expected under such high stresses,
but the substantial recovery suggests a mechanism in which considerable elastic back-stress must be present in ligaments which have
deformed into porosity, and an excellent memory of the original
structure is retained. Under uniaxial stresses, this memory is
present but thermal activation is required to remove the permanent
sets. The significant extent of microfracturing under uniaxial
deformation has already been discussed, and a similar contribution
to deformation under multiaxial stresses is probable. The reversible
nature of the process warrants further study, but the work indicates
considerable strength, even in porous graphite, under equitriaxial
compression.

Combining Fig. 3 for the biaxial envelope with Fig. 5 for the
failure locus under triaxial stresses (in the $\sigma_1 = \sigma_2$ plane) gives
an indication of the complete three-dimensional fracture envelope
for graphite, although there remain some regions which have not
received experimental study and others only examined to a limited
extent. Figure 6 illustrates the form this three-dimensional envelope would take for an isotropic material, having its axis along
the equitriaxial stress locus.

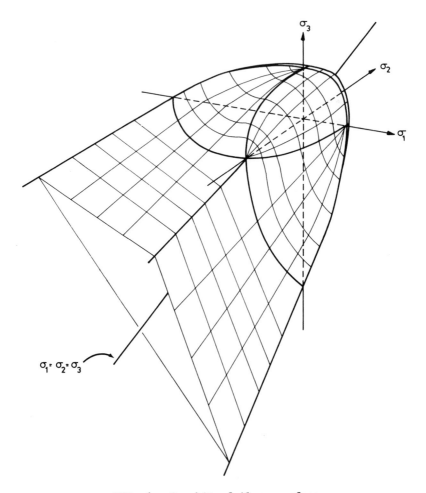

FIG. 6. Graphite failure surface.

V. STATISTICS OF FRACTURE, SIZE EFFECTS,
AND NONUNIFORM STRESS CONDITIONS

The previous section described the extent of knowledge of the failure
surface of polycrystalline graphite in terms of mean stresses but
pointed out the need for a statistical description. The discussion
also concerned itself almost entirely with near-uniform stress fields,
and one of the most difficult problems is the definition of a failure

criterion under a nonuniform stress condition. Before any recom-
mendations are made in design, the materials scientist must satisfy
himself that he can interrelate fracture data from different laboratory tests.

The Griffith theory [39] that fracture of brittle materials is
initiated by the inherent flaws in the material leads to statistical
theories in which the probability of failure of a sample is determined by the distribution of flaws and the stress pattern. The
model of brittle failure commonly adopted is the weakest link hypothesis in which failure occurs when the stress intensity at any one
flaw reaches the critical value for crack propagation, and in samples
selected from a parent population, it is the distribution of extreme
values that is statistically important. The consequences of this
model are:

1. Variability in strength test data requiring a specified
    acceptable failure probability for design purposes.

2. The prediction of a size effect, since specimens of large
    volume have a greater chance of containing a large flaw and
    are therefore weaker.

3. Under nonuniform stress conditions the volume under maximum
    stress is relatively small and so is the probability of
    finding a large flaw within this volume. Hence maximum
    stresses at failure (for example in a bend test) can exceed
    those obtained in specimens of similar volume subjected to
    uniform stress. In addition, failure may not initiate at
    the point of maximum stress, since the combination of
    stress distribution and severe flaw distribution must be
    considered.

The statistical analysis most often employed in fracture problems is that due to Weibull [59]. It can be shown that the probability of fracture of a material composed of a series of independent
elements under stress σ is of the form:

$$S(\sigma) = 1 - e^{-B}$$

where B is known as the risk of rupture, and for a sample of volume
V, was chosen by Weibull to have the form:

$$B = \int_v \left(\frac{\sigma - \sigma_u}{\sigma_0}\right)^m dV$$

where

σ = tensile stress in the element dV
σ_u = stress below which the failure probability is zero
σ_0 = normalizing parameter
m = homogeneity factor characteristic of the material (high values of m indicate a greater uniformity in the crack distribution)

The parameters m, σ_u, σ_0 may be obtained from experimental data by best-fit methods. In practice, σ_u is difficult to define independently with any precision, and for different graphites, it has been observed that almost equally good fits to the data can be obtained with a range of appropriate values for σ_u and m, including usually σ_u = 0, hence the latter value is generally adopted.

The theory can in principle be used to predict failure probabilities for components differing in volume and applied stress state, including stress gradient and multiaxial stresses, and has been tested by intercomparison of different types of laboratory test specimens. Equal probabilities of failure apply under different situations when the risk of rupture is the same, and hence comparisons are made by equating the relevant values of B. For example, in uniform tension,

$$B = V_t \left(\frac{\sigma_t}{\sigma_0}\right)^m$$

and in simple bend,

$$B = \frac{V_b}{2(m + 1)} \left(\frac{\sigma_b}{\sigma_0}\right)^m$$

where V_t, V_b and σ_t, σ_b are the appropriate specimen volumes and maximum tensile stresses, respectively. Equating the values of B gives

$$\frac{\sigma_b}{\sigma_t} = [2(m + 1)]^{1/m} \left(\frac{V_t}{V_b}\right)^{1/m} \tag{5}$$

and for different volumes V_1, V_2 of the same test sample

$$\frac{\sigma_1}{\sigma_2} = \left(\frac{V_2}{V_1}\right)^{1/m} \tag{6}$$

Price and Cobb [71] have tested the Weibull theory with results
of tensile tests, bend tests, and thick-walled tube burst tests (with
and without a superimposed axial stress) on H-327 graphite. Generalizing Eq. (5) for each test, they express the ratio of the median
strength σ_{nu} of samples with these nonuniform stress distributions
to that of uniform tensile specimens, σ_t, to be

$$\frac{\sigma_{nu}}{\sigma_t} = \left(\frac{V_t}{V_{nu}}\right)^{1/m} \quad \text{(stress distribution term)} \tag{7}$$

Bend test results did not show the predicted dependence on sample
width, and strengths predicted from the uniform tensile data (m = 8)
were higher than observed values (with the bend sample volume lower
than the tensile samples). For tube-burst and biaxial tube-burst
tests, the predicted values were lower than observed data (sample
size higher than the tensile samples). When the volume term in
Eq. (7) was placed equal to unity, however, and the stress distribution term alone used, better agreement was obtained. They concluded
that changes in failure probability due to nonuniform stress distributions predicted by the Weibull theory are supported by the observations on H327 graphite, but the expected dependence on sample volume is not observed.

Brocklehurst and Darby [76] have also applied the Weibull theory
to data on near-isotropic graphites for uniaxial tension, bend, and
internal pressure tests on ring specimens as part of an attempt to
interrelate these different types of tensile test with a common
failure criterion. It was observed, in general, that in all tests
the maximum stress at failure as calculated by elastic theory

exceeded that under uniform tension and increased with the degree of
severity of the stress gradient. The Weibull theory was applied to
the different tests but failed to give consistent values, fitting
all the data available, for the so-called material parameters, in
particular the Weibull modulus m. Over two orders of magnitude in
specimen volume, there was little dependence of both bend and tensile strength on volume, requiring a high m value (∿18) according
to Eq. (6), and this was consistent with the m value describing the
distribution of bend strength at constant volume, but the ratio of
bend to tensile strength (factor ≈1.6) required a much lower value
m ≈ 6 according to Eq. (5). It must be remembered, of course, that
the theory strictly applies to brittle, linear-elastic materials,
and graphite shows nonlinearity in its stress-strain behavior. However, attempts to analyze the influence of this nonlinearity on the
maximum stresses [76] indicated that the effect was probably small.

Amesz et al. [77] compare 4-point bend and tensile data for a
range of commercial graphites differing in maximum grain size from
0.4 to 6.7 mm. They show that a correction to the bend strength
results for nonlinearity in the stress-strain curve (using Wooley's
[78] exponential relation) reduces the bend/tensile strength ratio
by about 10%, typically from 1.45 to about 1.3, and this ratio is
substantially independent of grain size. Weibull m values obtained
from tensile data range from about 7.5 to 11.5. These values in
Eq. (5) with bend specimen volumes about four times greater than
those of tensile specimens showed theoretical strength ratios considerably lower than the experimentally observed values, and they
also concluded that the Weibull theory overestimates the volume
effect, giving pessimistic results when extrapolating experimental
data to higher volumes. Thus attempts to relate strengths from
different laboratory tests using a statistical Weibull treatment
have all concluded that the relationships predict a greater dependence of strength on specimen volume than is experimentally observed.

Lungagnani and Krefeld [79] show an increasing mean tensile
strength with decreasing specimen volume supporting Eq. (6), but
the volume dependence appears to require m ≈ 9, while they quote an

m derived from the distribution at constant volume of 3.4. This
latter value, however, is associated with a positive value of σ_u and
could be significantly higher if this parameter were assumed to be
zero.

In an earlier study on a British reactor grade graphite, Mason
[50] found that small-bend specimens showed the reverse effect of
an increasing strength with increasing volume, passing through a
peak as larger specimens were tested, to give the decreasing strength
with further volume increase characteristic of a weak-link material.
He attributed the size effect for the small specimens to the coarse-
grained structure of the material relative to the specimen size
(about 0.2-in. minimum depth). He pointed out that the weak-link
theory assumes that the rupture of any small element of volume dV
is not influenced by the presence of its neighbors, which is not true
if the size of the element is not large relative to the size of
defect initiating the rupture. Hence when specimen dimensions are
not large compared with the inherent defect size, a significant
portion of the specimen must already be regarded as having failed,
and the assumptions of the weak-link theory are not valid under
these circumstances.

Lanza and Burg [80] studied a wide range of graphites under
tension and compression to examine the effect of specimen volume.
They also observed a rise in strength with increasing volume, passing through a maximum for most materials at a particular volume,
and then decreasing for further increase in specimen size. Data
in compression showed that the critical volume at which the maximum
strength was observed increased with increasing maximum grain size
and also with impregnation (e.g., in impregnated material, the relationship indicated a critical volume increase from 1 to 7 cm^3 with
change in grain size from 0.5 to 5 mm).

These experiments show that the influence of grain size relative to specimen dimensions must be considered when obtaining
strength data on different materials and also when attempting to
interpret volume effects by the statistical Weibull treatment,
when the influence of grain size must be isolated from a true

weak-link size effect. To investigate these effects in both 4-point
bend and tension on the same material, a detailed study of the effect
of specimen geometry and size has been performed on both these modes
on specimens of IMI-24 graphite. One sample of each geometry was
machined from each of six sections cut from one graphite block.

Figure 7 shows the tensile and bend strength as a function of
the volume of the cylindrical tensile specimen and the rectangular
beam inner span, respectively, each point representing the mean and
standard deviation of six specimens. The bend strength data for
specimen volumes >1 cm^3 show a decreasing strength with increasing
volume and are approximately consistent with the predicted dependence (shown dotted) from Eq. (6) and a Weibull m value of 16. This
value of m = 16 is also obtained from the distribution of 30 bend
strengths at a constant volume of 25 cm^3 shown in Fig. 8. At
specimen volumes <1 cm^3, however, the bend strengths depart from the
predicted curve in Fig. 7 and show a tendency to decrease with specimen volume, attributed by Mason in another graphite to an effect
of grain size relative to specimen dimensions.

The tensile data in Fig. 7 show an increasing strength with increasing volume, tending to saturate at a maximum value for volumes
>8 cm^3, suggesting that the grain size effect is dominant and much
more pronounced in specimens subjected to uniform stress than when a
stress gradient is present. For a grain size effect in tensile
specimens, the diameter should be the important geometric variable,
and Fig. 9 shows that the strength falls rapidly for specimen diameters <10 mm. In bend specimens on the other hand, where stresses
change very significantly over the length of a grain and the effect
is much less pronounced, the bend strength is almost constant as
specimen thickness is reduced due to a balance between the opposing
effects of grain size and the weak-link theory. Table 3 gives some
examples of the apparent lack of sensitivity to bend specimen
thickness when the specimen volume is small. The outer and inner
span dimensions were 38 and 19 mm, respectively.

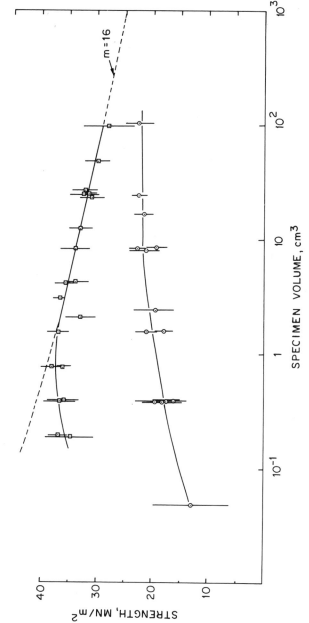

FIG. 7. Effect of specimen volume on 4-point bend and tensile strength of IM1-24 graphite.

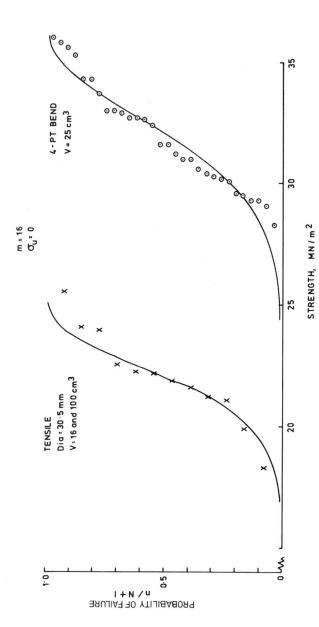

FIG. 8. Distribution of 4-point bend strength at constant volume and of tensile strength for specimens of maximum diameter for IM1-24 graphite.

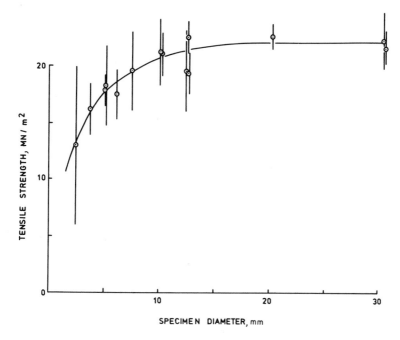

FIG. 9.   Effect of specimen diameter on tensile strength of
IM1-24 graphite.

TABLE 3

Some 4-point Bend Strength Data at Small Specimen Sizes
for IM1-24 Graphite

Width (mm)	Thickness (mm)		
	6.3	3.2	1.6
	Strength (MN/m^2) with SEM for six specimens		
12.7	37.0 ± 0.7	36.2 ± 0.6	36.8 ± 1.2
6.3	38.1 ± 0.9	36.0 ± 1.2	37.0 ± 0.7
3.2		34.9 ± 1.8	

Outside the range of small specimen effects in bend specimens, a decrease in strength is observed with increase in specimen volume at constant thickness (constant stress gradient), but there is also a small decrease in strength when the specimen thickness is increased (decreasing stress gradient) at constant volume. Table 4 illustrates this point, showing the mean strength (and standard error of the mean for six specimens) as thickness h is changed at constant specimen volume (by increasing breadth to compensate) and as volume is changed at constant thickness (constant stress gradient). This implies that stress gradient relative to grain size may be playing an additional role to the normal Weibull effect, for which the mathematical representation considers the distribution of flaw strengths but does not take account of a real defect size.

It is clear from the previous discussion that the distribution of tensile strength at constant volume must not be analyzed when the data include variations due to different specimen diameters, even though variations in specimen length compensate, since low m-values will result from the scatter introduced by the diameter variations.

TABLE 4

4-point Bend Strength Data, IM1-24 Graphite, for Different
Thicknesses at Constant Volume and for Different Volumes
at Constant Thickness

| Volume V(cm^3) | Thickness h (mm) | | | | |
	50.8	25.4	12.7	6.3	3.2
	Strength (MN/m^2) with SEM for six specimens				
98.5			28.1 ± 1.8		
24.6	30.5	31.0 ± 1.0	31.6 ± 0.8	32.1 ± 1.0	33.7
8.2			34.1 ± 1.1		
4.1			33.8 ± 0.9		
3.07			36.7 ± 0.4		

There are insufficient data at high specimen volumes to satisfy this requirement and give an independent Weibull analysis. However, the two sets of tensile strength data from the specimens of the largest diameter (30 mm) have been combined in Fig. 8, ignoring the factor of six difference in volume between the two sets. The probability of failure distribution is compared with a curve based on $m = 16$, which is seen to give a reasonable description of the data, even though the volume difference between the two sets would tend to increase scatter and decrease the m value. Thus while the parameter is not determined independently, the tensile strength distribution of the largest diameter specimens appears to be consistent with the Weibull modulus found from both the volume dependence and the distribution at constant volume of the bend specimens.

In speculating on the volume dependence at specimen volumes $>10^2$ cm^3 in Fig. 7, two possibilities appear likely. Either both bend and tensile strengths will decrease with volume according to the weak-link model, or they will tend to remain constant if there is a finite cut-off in maximum flaw size, and this flaw is always sampled above a certain specimen volume. In the former case, the laboratory tests on tensile samples with diameters in the range 10 to 30 mm give the same strength result because the balance between weak-link and grain size effects is occurring at much higher specimen volumes than in the nonuniform stress condition. In the latter, the data represent a "true" tensile strength for the material. In either of these alternatives, the bend/tensile strength ratio could remain constant at high specimen volumes.

Figure 7 shows that the experimentally determined bend/tensile strength ratio is not unique but varies from about two at low specimen volumes to about 1.3 at the higher volumes ($\sim 10^2$ cm^3). If a true limiting value applies when the grain size effect is no longer important in tension, this experimental value at high specimen volumes approaches the theoretical ratio of 1.25 given by Eq. (5) with $m = 16$, particularly if a small correction applies to the bend strength for nonlinearity in the stress-strain curve.

Thus apparent anomalies in the original interpretation of some
of these data [76] can be explained by the recognition of a grain
size effect and its interaction with the weak-link effect when the
latter could quantitatively describe, with consistent values of the
Weibull parameters,

1. The distribution of both bend and tensile strength at con-
    stant volume.
2. The effect of specimen volume (at least in bend tests).
3. The bend/tensile strength ratio.

We have already seen that the Weibull approach can be used to give
a reasonable description of experimental data under biaxial tension,
and in principle, the approach can be used to predict the failure
of components under complex stress states by using finite element
techniques.

Thus there are several features of graphite fracture behavior
under tensile stresses that are explicable on a weak-link model, but
it is clear that a complete description must also take into account
the relationship between grain size and specimen size, which is less
important as stresses become less uniform, and an effective statistical approach should be sought which can accommodate both the
grain-size and weak-link effects.

Finally, a further important practical example of nonuniform
stress conditions arises when failure occurs under thermally induced stresses, posing similar problems in relating the failure
condition to conventional strength data. Powell and Massier [81],
for example, succeeded in causing failure in graphite specimens
subjected to thermal stress, and defined heat flux rupture limits
for different graphites, but had difficulty relating the calculated
thermal stress at failure to the tensile strength determined in
standard tests. Andrae [82] has proposed a simple criterion for the
onset of fracture initiation under nonuniform stresses, based on
equating the average strain energy for the anticipated initial fracture surface in the test specimen to that required in a standard
tensile test. This procedure gave good agreement with experiment

in a number of tests which included thermal stress fracture in thin annular discs. Difficulties encountered in applying this type of approach in other tests involving stress gradients [76] are related to defining the area over which the averaging should be performed.

VI. EFFECTIVE WORK OF FRACTURE, FRACTURE TOUGHNESS, AND INHERENT DEFECT SIZE

In recent years, there has been a growing interest in the application of fracture mechanics techniques to graphite, and the resistance to crack propagation, or fracture toughness, has been investigated by several authors using different techniques on a variety of materials.

Fracture of the material is defined by critical values of either energy or stress parameters. Based originally on the Griffith [39] approach, the strain energy release per unit area of new fracture surface formed is identified with the effective surface energy γ_c. The critical value of strain energy release rate at fracture is frequently defined in terms of the area of material fractured by the fracture toughness parameter $G_c = 2\gamma_c$. Alternatively, a stress intensity factor K is derived which has dimensions (stress) x (length)$^{1/2}$ and defines the stress distribution at the tip of the critical crack. The magnitude of K depends on the structural geometry and the loading system and, in principle, for a given structure containing a crack, a knowledge of the critical value for the material K_c and the computed value of K at the crack tip indicates whether or not failure will occur in that component.

Thus both G_c and K_c are used in the literature as fracture toughness parameters, and the relationship between them is given by:

$$K_c^2 = EG_c \quad \text{for plane stress}$$

or   (8)

$$K_c^2 = \frac{EG_c}{1 - \nu^2} \quad \text{for plane strain}$$

where

    E = Young's modulus
    ν = Poisson's ratio

Fracture toughness is a minimum under plane strain conditions when
the mode is a wedge opening of the crack (mode I) and the critical
value under these conditions K_{Ic} is generally regarded as the con-
trolling parameter in design. In practice, ν is small for most
graphites, ≈ 0.2, and hence the relationship $K_{Ic}^2 = EG_{Ic}$ applies approximately in this condition also. For laboratory test specimens
of width W, the relationship between K_c, applied stress σ, and flaw
size a, at fracture is of the form:

$$K_c = \sigma(\pi a)^{\frac{1}{2}} f\left(\frac{a}{W}\right) \tag{9}$$

where f(a/W) is a geometrical correction factor. Thus for a given
geometry and loading system, $\sigma(a)^{\frac{1}{2}} = $ constant.

Working with beam specimens notched in such a way as to give
controlled crack growth, Tattersall and Tappin [83] measured the
effective work of fracture of a reactor graphite to be $\sim 10^5$ erg/cm^2,
$(10^2$ J/m$^2)$.

Davidge and Tappin [84] distinguish between the effective surface energy at crack initiation γ_i, which may be obtained by an
analytical method (load criterion) or an experimental compliance
method, and the work of fracture γ_f obtained by measuring the work
done to fracture a specimen completely in a controlled manner,
averaged over the complete sample fracture process. For graphite,
$\gamma_i < \gamma_f$ due to the tortuous crack path and high-energy absorption
by secondary cracking. They showed that for British PGA graphite
the analytical and compliance techniques using bend tests gave good
agreement for γ_i with no overall systematic variation with notch
depth in the specimens tested. On the other hand, work of fracture
determinations of γ_f gave decreasing values with increasing notch
depth, but in all tests $\gamma_f > \gamma_i$, for which a value ≈ 50 J/m^2 was
obtained. Values of γ_f were in the range 1 to 2 x 10^2 J/m^2 with
decreasing notch depth. Thus crack propagation is more difficult
than crack initiation.

Vitovec [85] has discussed the difference between G_f and G_i in terms of the work required for slow crack propagation, using compact tension specimens and bend specimens under controlled crack growth conditions. He obtained K_c values $\approx 10^3$ lb/in^2 · in$^{\frac{1}{2}}$ (≈ 1.1 MN/m^2 · m$^{\frac{1}{2}}$) for an extruded nuclear graphite RC4 (Airco-Spear).

The temperature dependence of the work of fracture G_f has been determined by Udovskii et al. [86] in notched bend tests on a graphite designated MPG-6. The values were sensibly constant for notch depths in the range a/W = 0.2 to 0.4. At 2000 to 2300°C, they observed an increase over the room temperature value of 10^2 J/m^2 by an order of magnitude, attributed to an increase in the dissipation energy by plastic deformation at the crack tip.

Corum [87] found in notched bend tests on EGCR-type AGOT graphite that G_i decreased at small notch depths, and he suggested this effect was caused by inherent flaws which effectively increased the artificial notch by a constant amount, estimated from tests on unnotched beams. He then found more consistent values. For this material, he obtained G_c = 0.39 in.lb/in^2 (68 J/m^2) parallel to extrusion and 0.29 in.lb/in^2 (51 J/m^2) perpendicular to the extrusion axis.

Yahr and Valachovic [88] investigated the effect of specimen geometry on the stress intensity factor at the onset of rapid crack propagation, K_c, for two different graphites, ATJ and AXM (POCO). They studied the effect of geometric variables including notch width, depth, and sharpness on notched bend specimens, and then compared values from other types of specimen including circumferentially notched tensile and compact tension types. In the bend tests, they also observed low K_c-values at small notch depths, the effect being less marked in the AXM material. They concluded that beams gave valid results if the machined notches were no wider than 0.006 in. and of sufficient depth to avoid the reduced values obtained at small a/W. The AXM material was isotropic with a mean value for K of 1.17 10^3 lb/in^2 · in$^{\frac{1}{2}}$ (1.3 MN/m^2 · m$^{\frac{1}{2}}$), and the ATJ graphite was more anisotropic with values in the range (0.7 to 0.8) 10^3 lb/in^2 · in$^{\frac{1}{2}}$, (0.8 to 0.9 MN/m^2 · m$^{\frac{1}{2}}$). For other specimen geometries, the tensile specimens of ATJ graphite tended to give

low values, and they concluded that notched beams and compact tension specimens were good geometries for fracture toughness determinations on graphite.

The application of fracture mechanics to general fracture problems in graphite will require the use of experimentally determined
K_c (or G_c) values on components containing the inherent random defects and flaws of the material rather than machined artificial
defects. Using their K_c-values with the strengths of the unnotched
beams, Yahr and Valachovic calculated effective values of the inherent notch depths of 0.010 in. for ATJ and 0.003 in. for AXM, and
concluded that they corresponded to the size of the largest natural
inhomogeneities in these materials determined by microscopic examination. Yahr et al. [89] subsequently used these data to predict the
failure of disc specimens subjected to a diametral compressive load,
and while demonstrating good agreement with experiment for the isotropic AXM material, the predictions were less precise for the more
anisotropic graphite ATJ.

A similar investigation of the effect of specimen type and geometry has been performed on a fine-grained version of graphite IM1-24
using specimens of the type shown in Fig. 10. Stress intensity values
were derived from the load criterion using equations given by Srawley
and Brown [90] for the bend and tensile tests and by Wessel [91] for
the compact tension specimens. These equations are all essentially
of the form given by Eq. (9) with different geometric factors.

The results may be summarized as follows. With edge-notched
beam specimens of depth 6.4 mm and width 19 mm, 3-point bend tests
gave K_c-values of 1.4 to 1.5 $MN/m^2 \cdot m^{\frac{1}{2}}$ with no significant difference in values from specimens with slot widths of 0.005 to 0.010 in.,
slot depths of $a/W = 0.1$ to 0.5 in., and variations in the specimen
outer support distance of 76 to 305 mm. Tensile tests on similar
specimens, and also compact tension specimens (with $W = 11$ mm,
$D = 6.4$ mm, $a = 5$ mm, slot width $= 0.125$ mm, in Fig. 10) gave values
of 1.3 $MN/m^2 \cdot m^{\frac{1}{2}}$ in reasonable agreement with the bend test results.
However, the circumferentially slotted round bar gave considerably
lower K_c-values of about 0.8 $MN/m^2 \cdot m^{\frac{1}{2}}$. Using precracked specimens

3 - POINT BEND

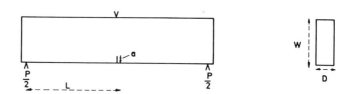

SINGLE EDGE NOTCHED TENSILE

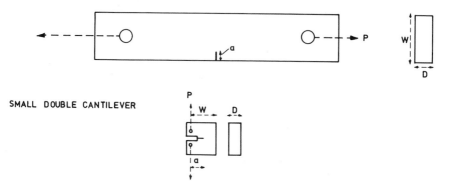

SMALL DOUBLE CANTILEVER

CIRCUMFERENTIAL NOTCHED ROUND BAR

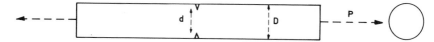

FIG. 10. Specimens used in work of fracture tests.

of the compact tension type, made from similar material, Marshall
and Priddle [92] obtained a value of $K_c = 1.5$ MN/m$^2 \cdot$ m$^{\frac{1}{2}}$, which is
in agreement with the foregoing range of results excluding those
from the circumferentially slotted round bars. Assuming unnotched
beams may be represented with an inherent edge flaw, the size of
this effective inherent defect is estimated from the range of K_c-
values and bend strength data to be in the range 0.25 to 0.4 mm,
which is small compared to the size of the machined notches but comparable to the maximum grain size of this graphite of 0.3 mm.
Vitovec and Stachurski [93] used the Tattersall and Tappin type bend
specimens to study certain fracture characteristics and their anisotropy for both extruded Speer graphite type RC4 and molded POCO

graphite type AXF-Q1. Averaged work of fracture values in air at
room temperature were 200 to 240 J/m^2 for RC4 graphite, increasing
with specimen size, and about 150 J/m^2 for the small AXF-Q1 specimens. The AXF-Q1 graphite is stronger than RC4 but more brittle.
Calculation of critical defect size gave 0.8 mm for RC4, which is
similar to the maximum size of the filler particles and not much
different from the pore sizes. For AXF-Q1 graphite, the calculated
defect size was 0.3 mm, which is considerably larger than the natural grain or pore size.

Stevens [94] also studied the fracture behavior of the fine-
grained isotropic POCO grade graphite AXF-Q1. Fracture surfaces
were examined using replica and scanning electron microscopy, and
also by transmission electron microscopy on specimens thinned by
ion bombardment. Groups of basal plane microcracks in the size
range 10^{-5} to 10^{-4} cm were observed, some of which linked to form
larger cracks. The pore size of the material was about 1 µm, and
the grain size, similar. Work of fracture determinations gave values
of about 22 ± 2 J/m^2 with double cantilever beam specimens and
70 ± 2 J/m^2 with notched beam specimens (of Tattersall and Tappin
type). The lower value of the former type of specimen was attributed to the way in which the fracture was confined to a single rather
than multiple crack, and when used with the bend strength of
90 ± 5 MN/m^2 and a modulus $\sim 10^4$ MN/m^2 gives a critical crack size
~ 0.01 mm, which is an order of magnitude greater than the observed
microcracks and at least four times greater than the largest pores.

A fracture mechanics approach to graphite failure problems involves the concept of an effective flaw size, and we have seen how
sizes derived from applying fracture mechanics results to tests on
unnotched beams have been identified with observable microstructural
features, such as maximum grain or pore size, with the possible
exception of POCO materials. The following experiment illustrates
an independent derivation of this parameter from laboratory tests.

Single edge-notched beams, 6.5-mm wide x 13-mm deep of graphite
grades SM2-24 and VNMC were broken in (1) tension (50-mm long) and
(2) 4-point bend (117-mm long with outer/inner knife edge distances

of 101/32 mm, respectively). The centrally located edge notch was
0.25-mm wide with varying depths up to 5 mm (a/W = 5/13). The mean
tensile strength and standard deviation for 12 unnotched specimens
was 12.0 ± 1.3 MN/m^2 for SM2-24 and 10.6 ± 1.5 MN/m^2 for VNMC graphite. The mean bend strength and standard deviation for 12 unnotched
specimens was 18.9 ± 0.8 MN/m^2 for SM2-24 and 18.8 ± 1.3 MN/m^2 for
VNMC graphite.

Six specimens were tested at each notch depth, and Figs. 11 and
12 show the variation in mean breaking load (normalized to unity for
zero notch depth), with the standard deviation, as a function of
notch depth in both bend and tensile tests, and also the number of
failures occurring at the notch. In Figs. 11(a) and 12(a), the load
to failure falls with increasing depth of artificial notch after an
initial period of relatively small or insignificant effect at small
notch depths. During this period, the point of failure is not necessarily at the notch, but as Figs. 11(b) and 12(b) show, this location becomes more probable as the notch depth increases until all
failures occur at the notch and the load to failure then subsequently
decreases markedly with further increase in notch depth. Thus
Figs. 11 and 12 define the notch depth which has little effect on
strength by measuring the load to failure and, in some ways more
informatively, by recording the number of fractures at the notch.
This notch depth may thus be interpreted as an effective maximum
inherent defect size for the material.

Alternatively, an effective flaw size may be determined as in
the earlier discussion from the relationship between the critical
stress intensity factor K_c and the stress at failure of unnotched
specimens. An evaluation of K_c by the load criterion for the
specimens containing the deeper notches gave values of 0.80 MN/m$^{3/2}$
for bend specimens and 0.85 MN/m$^{3/2}$ for tensile specimens of SM2-24
graphite. Corresponding values for VNMC graphite were 0.92 MN/m$^{3/2}$
for bend specimens and 0.91 MN/m$^{3/2}$ for tensile specimens. These
toughness values and the respective strengths of specimens with no
artificial flaw are used with the appropriate relationships to estimate a value for the effective inherent flaw, giving for SM2-24

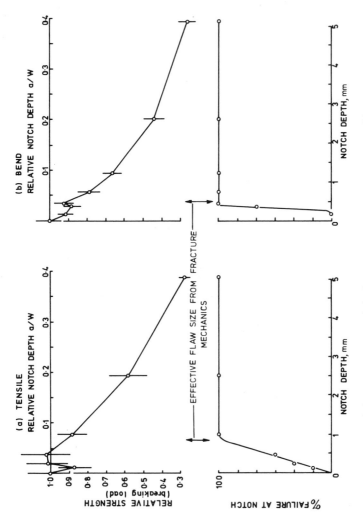

FIG. 11. (a) Tensile and (b) 4-point bend tests on SM2-24 graphite beams showing influence of edge-notch depth on breaking load and incidence of failure at the notch, compared with estimate of flaw size from fracture mechanics.

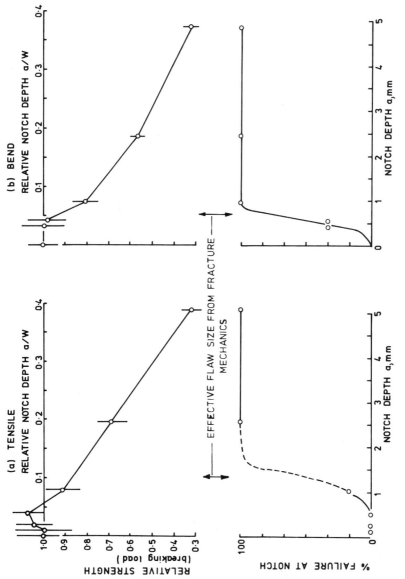

FIG. 12. (a) Tensile and (b) 4-point bend tests on VNMC graphite beams showing influence of edge-notch depth on breaking load and incidence of failure at the notch, compared with estimate of flaw size from fracture mechanics.

graphite about 0.5 mm in bend, 0.8 mm in tension, and for VNMC material 0.7 mm in bend, 1.3 mm in tension. These values of effective
inherent flaw size are illustrated by the arrows in Figs. 11 and 12
and show good agreement with the independent experimental determina-
tion of the largest notch depth which does not significantly decrease
the strength of the specimens. It should be noted however that
while self-consistent in themselves, the bend and tensile data are
different. The tensile tests give values for the effective flaw
size which are about twice those obtained from the bend tests.

Table 5 summarizes the values reported here and in the literature for the fracture toughness parameters and effective work of fracture of a wide range of graphites with approximate typical values
where available for the Young's modulus and strength, and evalua-
tions of effective inherent flaw size. The concept of using fracture mechanics rather than a maximum principal stress criterion is
in particular circumstances rather academic since these parameters
are related through an effective inherent flaw size, and the variations in strength are due in part to variations in this natural
defect size. It is, however, its potential value in defining the
failure condition under general component geometry and loading conditions that has led to fracture mechanics studies on graphite. In
such general situations, it would be necessary to define the critical
value of the fracture toughness parameter and the effective inherent
flaw size of the material.

Finite element techniques may be used to evaluate the rate of
release of strain energy with the progression of this crack under
complex geometry and loading conditions [95]. This is achieved by
calculation of the strain energy as successive nodes in the finite
element mesh are removed, although it is necessary to define the
direction in which the crack will propagate. In principle, therefore, this exercise may be performed to determine the failure condition for a given structure by comparison with the effective critical value of the fracture toughness parameter for the material.

It must be emphasized again that any proposed failure criterion
must be proved in relatively simple laboratory tests by the materials

TABLE 5

Effective Work of Fracture, Fracture Toughness, and Inherent Defect Size Data

Reference	Graphite	Test	K_c $(MN/m^2 \cdot m^{1/2})$	G_c (J/m^2) $\times 10^2$	Effective work of fracture G_c (J/m^2) $\times 10^2$	Effective inherent defect size (mm)	Young's modulus E (MN/m^2) $\times 10^3$	Strength tensile bend (MN/m^2) σ_t	σ_b
Tattersall and Tappin [83]	PGA	bend		1	1				
Davidge and Tappin [84]	AGOT	bend ‖		0.7	2-4	}1-1.5	11	10	14
Corum [87]		(bend ⊥)		0.5		}			
		(bend ⊥)							
Yahr and Valachovic [88]	ATJ	(compact tension)	0.8-0.9			0.25			
Vitovec [85]	AXM	compact tension	1.3			0.07			
Vitovec and Stachurski [93]	RC4	compact tension	1.1						
	RC4	bend RT air			2.0-2.4	0.8	10	12	23
		bend RT vac			3.4				
		bend 750°C vac			3.25				
	AXF-Q1	bend RT air			1.6	0.3		33	96
		bend RT vac			1.9				
Stevens [94]	AXF-Q1	bend			0.7				90
		DCB			0.2				
Udovskii et al. [86]	MPG-6	bend (a/w ∿0.25) RT			1	0.01			
		1000°C			3.4				
		2000°C			6				
		2300°C			10				
Marshall and Priddle [92]	reactor graphite	compact tension (bend ‖)	1.5						
This review	IM1-24	(compact tension)	1.3-1.5	1.4-1.9		0.25-0.4	12	25	39
	SM2-24	bend ‖	0.80	0.8		0.5	8	12	19
		tensile ‖[a]	0.85						
	VNMC	bend ‖	0.92	0.9		0.7	9	11	19
		tensile ‖[a]	0.91						

[a]On edge-notched beams.

scientist before being offered to the engineer for use in more complex situations. Limited tests of the method in predicting failure
conditions have been made and some success has been claimed [89].
Marshall and Priddle [96] have demonstrated that the relationship
between bend and tensile strength may be regarded as a consequence
of the difference in the respective functions for (a/W) in Eq. (9)
for a particular model of edge-notched beams tested in the two modes.
However, there are difficulties still to be resolved. The evidence
discussed previously (Figs. 11 and 12) indicates that stress gradients in the test result in smaller values of the effective inherent
flaw size than those obtained in uniform tension. This result is
believed to be related to the earlier observation that the effect
of grain size is less important in bend than in uniform tension (discussion of Fig. 7 in Sec. V). It is also noted that, in the exercise described previously when the inherent defect size was defined
at about 1 mm for two graphites by independent methods (including
fracture mechanics), the strength of beams reduced in thickness to
about 1 mm was no lower than that obtained on thicker specimens. In
reality, of course, the material is not two dimensional. The beam
has a width which is considerably greater than the inherent defect
size, and this is the reason for its maintained strength, but a
method must be found of taking these points into account in order to
specify a general failure condition.

VII. FATIGUE

Graphite subjected to alternating stresses, at levels well below the
normal static strength of the material, will show fatigue failure.
This phenomenon is believed to be a result of microcrack growth and
is qualitatively consistent with current fatigue theory and the view
discussed earlier that there is a significant contribution from
plasticity at crack tips to the deformation of graphite. Most
experimental investigations have resulted in σ - log N curves obtained under particular stress patterns, usually either a mean stress
of zero (equal tensile-compressive stress amplitudes) or stress

cycled between zero stress and a tensile value. Experimental fatigue
data reflect the variability in graphite static strength and assuming
a similar worst flaw mechanism of failure, must also be similarly
dependent on the distribution of flaws and applied stress, giving
rise to consideration of the effects of specimen volume, modes of
stressing, and the general statistical aspects of the process.

Green [97] showed from reverse-band fatigue tests performed
both at room temperature and at 1950°C on grade AUF graphite (Union
Carbide Corp.) that this material has an apparent endurance limit
which increases with temperature. At room temperature, the limit
applied for $N > 10^4$ cycles and was defined at 17 MN/m^2 (first-cycle
strength not indicated), rising to 30 MN/m^2 at 1950°C (in He) with
an almost flat $\sigma - N$ curve. This behavior is consistent with the
increase in static strength with temperature shown by most graphites,
but the factorial increase in endurance limit was larger than that
indicated for the static strength of this material.

Barabanov et al. [98] examined an extruded anisotropic graphite
(density 1.8 to 1.9 g/cm^3) under rotating bending at $N > 10^3$ cycles.
Fatigue limits were about 14 MN/m^2 parallel and 10 MN/m^2 perpendicular to extrusion, corresponding, respectively, to about 0.5 and
0.4 times the mechanical strength of the material in the two directions.

Sato et al. [99] studied the fatigue behavior of four graphites
at room temperature for $N > 10^3$ cycles in rotary bending, tension-
compression, and torsion. Their results also tend toward a fatigue
limit, for $N > 10^5$ cycles in each mode of testing, in the region of
0.7 to 0.8 times the appropriate first cycle strength.

Leichter and Robinson [100] used reverse bending up to 5×10^8
cycles at room temperature in studies on grade EP-1924 graphite (a
fine-grained material, 0.001 in., also designed AXF, of density
1.88 g/cm^3). They expressed their data in terms of homologous stress,
i.e., the ratio of applied stress to the first cycle strength, and
found little difference in the resulting curves whether this ratio
was based on the "mate" strength of specimens cut adjacent to the
fatigue specimen, or on the average strength of the sample. A

Weibull statistical analysis was applied, and at a 1% probability
of failure, the endurance limit corresponded to a homologous stress
of about 0.47 and at 50% probability, about 0.65. The normalized
homologous stress represents the ratio of the operational stress
to the expected strength under the same test volume and stress system and hence tends to eliminate variables of geometry and loading
pattern. Leichter and Robinson combined their own and other data
from the literature to show that the use of this homologous stress
gives a good fatigue correlation for a range of different graphite
grades.

Wilkins [101] proposes an alternative approach to obtaining and
presenting dynamic fatigue data obtained under a constant applied
stress σ_a, illustrated by experimental data for extruded RC4 graphite
(Speer Carbon Co.) in 4-point bending, cycled between stresses
$\sigma_a/2 \pm \sigma_a/2$ after proof stressing to σ_a. Fatigue life is considered
to be a function of σ_h and σ_a where the homologous stress $\sigma_h = \sigma_a/\sigma_i$
and is based on a statistical estimate of the individual first
cycle strength of the fatigue specimen σ_i. This estimate is obtained
from the known strength distribution in the same test mode. It is
assumed that the weakest specimen fails first in fatigue, etc., in
order to match the distribution of fatigue lives with σ_i values.
Thus the fatigue test at a fixed value of σ_a readily defines a fatigue curve of σ_h V N, eliminating the scatter normally associated
with such data and allowing an estimate of the cumulative probability of fatigue failure. The reader is recommended to read the original paper for details of the statistical treatment, but the approach
offers the engineer a method of making a realistic assessment of the
probability of fatigue failure and, where practicable, of measuring
the increase in permissible operating stress of a component gained
by proof testing, against the component losses during the proof
test. Wilkins has also applied this approach to "static" fatigue [44].

As noted previously, the homologous stress approach offers the
possibility of generalizing fatigue data in terms of σ_h for different
specimen volumes and different mode-of-loading conditions, and also
possibly for different materials. Wilkins and Reich [102] extended

the experimental work with test data on POCO graphite in both biaxial loading (center loaded discs) and 3-point bending (also between stresses of $\sigma_a/2 \pm \sigma_a/2$ as for RC4 graphite) to indicate that, for practical purposes under these stress conditions, one conservative relationship between σ_h and N may be used to describe the behavior of the two types of graphite in these different modes of loading.

Figure 13 compares the curves given by Wilkins [101] for RC4 graphite under a homologous stress $\sigma_h = \sigma_a/\sigma_i$ and for two constant absolute values of σ_a, with data generated on grade IM1-24 graphites under a homologous tensile stress $S_h = \sigma_a/\bar{\sigma}_i$, where $\bar{\sigma}_i$ is the mean value of the first cycle (static) strength, and between zero stress and the peak value. The IM1-24 data thus reflect the total scatter due to the distributions of initial static strength and the effective crack growth rates of the individual specimens. The two curves given by Wilkins represent the effect obtained by proof testing to a higher stress and then fatigue testing the remaining population under this higher peak operational stress. The IM1-24 data were obtained in different testing modes, namely uniform tension, uni-axial bend, and biaxial bend center loaded discs. They include results on irradiated bend specimens whose absolute strength was increased by a factor of two on irradiation, and for which the stress-strain relation was initially more linear (Fig. 14). Normal-ized to the mean strength, the effect of testing mode (and of differ-ences in starting materials) is considerably reduced, and the com-bined data and their scatter are bracketed by the two curves given by Wilkins for RC4 graphite. The agreement is reasonable, although it must be noted that there are differences in the specified stress systems. For the IM1-24 data, increased endurance is obtained by decreasing the applied peak stress to any member of the population. For the RC4 material, increased endurance occurs at a fixed stress applied to stronger specimens in the population. Wilkins curves do not define a fatigue limit but clearly under these conditions of purely tensile stresses with a mean stress equal to half the peak applied stress, the mean endurance level is high for homologous stresses ≤ 0.65.

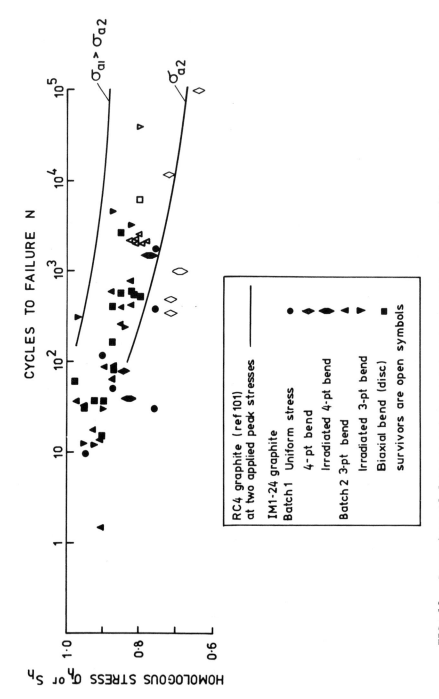

FIG. 13. Comparison of fatigue endurance data expressed by two definitions of homologous stress (see text), both involving stress cycles from zero to a peak tensile value.

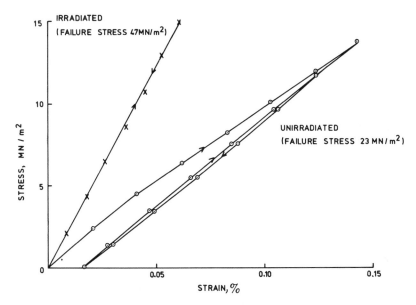

FIG. 14. Stress-strain behavior of IM1-24 graphite before and after fast-neutron irradiation to 10^{19} n/cm^2 at 50°C.

Marshall and Priddle [92] applied a fracture mechanics approach to the room temperature fatigue of reactor grade graphites, investigating the applicability of relationships applied to metal fatigue of a form in which the crack growth rate per cycle da/dN is proportional to $(\Delta K)^r$, where ΔK is the applied range of stress intensity factor and r is a constant. They used a compact tension type of specimen and measured the *macroscopic* growth rate da/dN over various ranges of ΔK for two graphite types, and found a relationship of the form:

$$\frac{da}{dN} = C(\Delta K - \Delta K_0)^4$$

where the critical value $\Delta K_0 = 0.85$ MN/m$^{3/2}$ is a minimum range for crack propagation which can be identified with da/dN values approximating the atomic spacing per cycle, and the constant $C = 1.36 \times 10^{-5}$ in these room temperature tests. The fourth power law is commonly found in metal fatigue studies and is identified as a measure of the extent of the plastic zone at the crack tip in that the volume of this zone is proportional to ΔK^4. This relationship therefore implies that plastic deformation is associated with fatigue.

Marshall and Priddle compared their predicted curve (which con-sidering their test must apply for the purely tensile stress range between zero stress and the peak value) with data (incorrectly attributed to Leichter and Robinson) obtained with a mean stress of zero under equal compressive and tensile stress amplitudes. The agreement was reasonably good but implies that the compressive cycle does not contribute to the damage and needs further comment. The fatigue limit was about half the static fracture stress, and this is predicted from the model since for the materials examined $\Delta K_0/K_c \sim 0.5$. The consequences and predictions using this model are worth further discussion and are examined. These authors examined and analyzed the particular situation when tensile stresses varied from zero to a maximum tensile value σ, with mean stresses $\sigma/2$. Their approach may be generalized to predict the influence of mean stress level and stress amplitude.

Assume the stress intensity range $\Delta K = \Delta\sigma(\pi a)^{\frac{1}{2}}$ provides the driving force for crack propagation, and the peak stress determines the extent of crack growth from a_i to a_f before ultimate failure occurs under the condition $K_c = \sigma(\pi a_f)^{\frac{1}{2}}$. The initial crack size a_i is determined by $K_c = \sigma_t(\pi a_i)^{\frac{1}{2}}$, where σ_t is the tensile strength (first-cycle fracture stress), and the model may also be used to predict the integrated crack growth from a_i to a_f as a function of the number of stress cycles N. Thus it can be shown that

$$CN\pi(\Delta\sigma)^2 = \frac{3K_c\Delta\sigma/\sigma_t - \Delta K_0}{3(K_c\Delta\sigma/\sigma_t - \Delta K_0)^3} - \frac{3\Delta\sigma(\pi a)^{\frac{1}{2}} - \Delta K_0}{3[\Delta\sigma(\pi a)^{\frac{1}{2}} - \Delta K_0]^3}$$

If $\Delta\sigma = B\sigma$ where B is a constant and $\sigma = K_c/(\pi a_f) K_c/\pi a_f^{\frac{1}{2}}$,

$$3K_c^2 CN\pi B^2\sigma_t^2\left(\frac{\sigma}{\sigma_t}\right)^2 = \frac{3B\sigma/\sigma_t - \Delta K_0/K_c}{(B\sigma/\sigma_t - \Delta K_0/K_c)^3} - \frac{3B(a/a_f)^{\frac{1}{2}} - \Delta K_0/K_c}{[B(a/a_f)^{\frac{1}{2}} - \Delta K_0/K_c]^3} \quad (10)$$

Since at fatigue fracture $a = f_f$ giving $(a/a_f)^{\frac{1}{2}} = 1$, $\sigma/\sigma_t = S_h$ may be derived as a function of $N \sigma_t^2$ for any value of B, which defines

$\Delta\sigma/\sigma$. Thus normalizing to a particular mean value σ_t^2, $S_h N$ curves defined by a constant relation of $\Delta\sigma$ to σ may be generated if ΔK_0 and K_c are known. Also, prior to fracture, Eq. (10) gives a/a_f as a function of N at any particular values of σ/σ_t and B, and since $a/a_i = (a/a_f)(a_f/a_i) = (a/a_f)/(\sigma/\sigma_t)^2$, the relative crack length a/a_i may be derived as a function of $N\sigma_t^2$. Each ratio a/a_i rises from unity as the cracks grow under stress cycling to a limiting value at fatigue fracture of $a_f/a_i = 1/(\sigma/\sigma_t)^2$. The model implies an inherent initial crack of effective size a_i. Whether or not it is a true representation of the absolute crack sizes involved is questionable, but the formal interrelationships are described by adopting this concept of a single effective crack.

Figure 15 shows $S_h N$ curves obtained from Eq. (10) for different values of B using Marshall and Priddle's data for ΔK_0 and K_c. The pair of curves for each B show the difference obtained using the property values for the two types of graphite examined. This difference between the two graphite types is smaller than the differences predicted by changing the value of $B = \Delta\sigma/\sigma$ and may explain why different materials tend to show experimentally similar homologous $S_h N$ curves. When $\Delta\sigma = \sigma$, $B = 1$, and the fatigue limit is $\Delta K_0/K_c$, which for these materials is in the range 0.50 to 0.57. As B is reduced from unity, the fatigue limit rises rapidly [= $(\Delta K_0/K_c)/B$], and for these materials, a mean stress level $\geq 0.75\ \sigma_t$ cannot produce fatigue because the maximum value of $\Delta\sigma$ ($\approx\sigma/2$) will not give a stress intensity range exceeding ΔK_0, and fracture will occur on the first cycle before any fatigue effects occur. Thus Fig. 15 considers only tensile stresses and shows that in round terms the fatigue limit increases rapidly as the mean stress rises from 0.5σ, and the higher mean stresses also produce greater differences between the two different graphites. (It should be noted that in each curve $\Delta\sigma/\sigma$ is constant and the mean stress σ_M falls with decreasing σ defined by $\sigma_M = \sigma - \Delta\sigma/2$. Only when $\Delta\sigma = 2\sigma$ is the mean stress constant at zero.)

The relative crack growth at each homologous stress σ/σ_t and stress range $\Delta\sigma$ may be derived from Eq. (10) and is illustrated

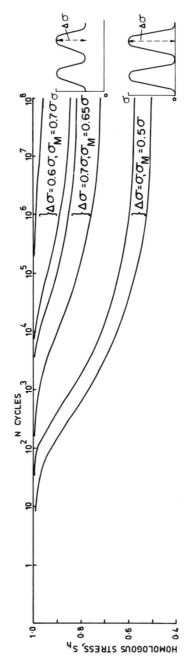

FIG. 15. Predicted fatigue endurance curves showing effect of stress amplitude and mean stress level (using Eq. (10) based on a model in Ref. 92).

for one of the graphites in Fig. 16 for the particular case $\Delta\sigma = \sigma$.
The crack growth as a function of N increases very rapidly to the
failure point and explains why it is difficult to detect large
strength deterioration in residual strength tests after stress
cycling.

Returning now to the comparison with experimental data, Fig. 17
shows data obtained on graphites similar to those examined by Marshall
and Priddle under the following stress conditions:

1. Mean stress = $\sigma/2$ ($\Delta\sigma = \sigma$) in the tensile region only.
2. Mean stress zero with equal tensile and compressive ampli-
    tudes.

The data were obtained under a variety of testing modes (uniform
tension, uniaxial, and biaxial bend), reduced to a common curve by
expressing in terms of the homologous stress S_h (normalized to the
mean value of the first cycle strength in the particular test mode),
and combined in Fig. 17 to distinguish only the effects of testing
under the stress conditions defined by (1) and (2). Within the
scatter of the data in Fig. 17 the differences are not very marked,
even though the specimens under stress condition (2) are exposed
to an additional compressive stress cycle of equal amplitude, but
the following points emerge. Under stress condition (2), the
data tend toward a fatigue limit of about 0.5 S_h, a limit in

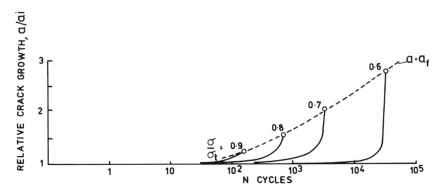

FIG. 16. Predicted fatigue crack growth as a function of endurance level at difference homologous stresses (for the particular case
$\Delta\sigma = \sigma$).

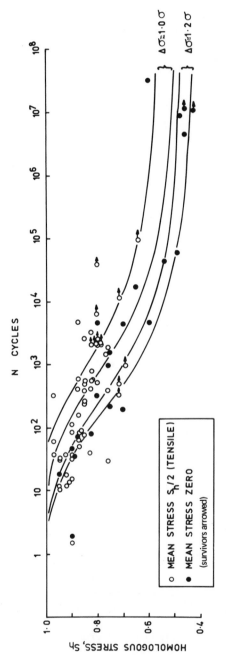

FIG. 17. Comparison of fatigue endurance data with (i) Mean stress equal to half the peak stress (tensile), and (ii) Mean stress zero.

approximate agreement with data on other materials as discussed
earlier. Under stress condition (1), data do not extend to high
N-values, but no specimens fatigued for S_h < 0.75, a large number
survived with S_h ≈ 0.8, and the mean curve tends to show marginally
greater endurance than under stress condition (2), although the
fatigue limit is not defined by the data.

The curves in Fig. 17 are derived as before from Eq. (10). The
prediction for $\Delta\sigma = 1.0\sigma$ may be considered to give a fair description
of the average data obtained under stress condition (1), although
it tends to predict a higher endurance at low N and is not adequately
tested at high N, but it does tend to overestimate the fatigue endurance of specimens subjected to stress condition (2). Thus the
results imply that the compressive stress cycle makes a small contribution to fatigue damage, presumably by the growth of a different
family of cracks which ultimately assist coincidental crack extension under a subsequent tensile stress. Figure 17 shows that a
better fit to the data obtained under condition (2) is given by the
prediction $\Delta\sigma = 1.2\sigma$, indicating that the compressive stress is only
20% as effective as the same tensile stress amplitude (N.B. a predicted curve for $\Delta\sigma = 2\sigma$, implying that tension and compression are
equally effective, grossly underestimates the fatigue endurance).
The homologous stresses are normalized to the tensile strength, and
from the earlier discussion on the relationship between tensile and
compressive strengths, one interpretation is that to produce the
same tensile stresses at the periphery of critical cracks, compressive stresses are only about 30% as effective as tensile stresses.
It is therefore not unreasonable that compressive stresses make a
smaller contribution to the effective crack growth in fatigue giving
rise to failure under a subsequent tensile stress cycle.

The difference between the fatigue endurance curves obtained
with a zero-mean stress condition (2) and those with a mean (tensile)
stress of $S_h/2$ is further illustrated in Fig. 18, using the data from
Fig. 17, which shows their respective distribution in endurance
levels (on a probability of failure basis) for homologous stress
values within the range $0.8 \le S_h \le 0.9$. The 50% failure probability
level corresponds to an endurance of 80 cycles with a zero-mean

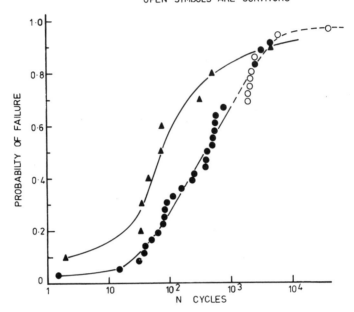

FIG. 18.   Probability of failure as a function of fatigue endurance for different mean stresses (peak homologous stresses limited to 0.8 to 0.9).

stress and about 400 cycles with a biased mean stress $\approx S_h/2$.

Figure 19 shows a similar distribution of fatigue lives for the
biased mean stress situation (mean stress $S_h/2$) at three different
groupings of homologous stress. At 10^3 cycles, the probability of
survival is about 70% for S_h = 0.75 to 0.80, 20% for S_h = 0.825
0.875, and <10% for S_h = 0.90 to 0.95. Surviving specimens are
included in both Figs. 18 and 19 and will tend to make the curves
pessimistic at high probabilities of failure.

In summary, the available data on dynamic fatigue in air at
room temperature indicate that the concept of a normalized homologous
stress tends to remove the differences in absolute endurance curves
obtained in different testing modes (under uniform, nonuniform, and
biaxial stress systems), and even possibly between different graphites.

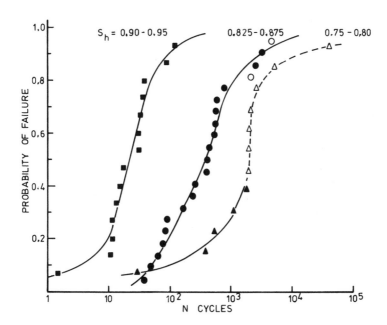

FIG. 19. Probability of failure as a function of fatigue endurance for different peak stresses (stress range zero to peak
values).

There is a difference between data obtained at zero-mean stress and
a tensile-biased mean stress, with the latter tending to show a
higher fatigue limit.

Under this purely tensile stress condition, Marshall and
Priddle's fracture mechanics approach predicts a mean fatigue limit
of 0.5 to 0.6 S_h (based on mean static strength) for particular
reactor grade materials, while the data do not show fatigue for
$S_h \leq 0.75$, supported also by Wilkins' data on other graphites.
Data obtained with a mean stress of zero at the same peak tensile
stress is somewhat more damaging. There is a lack of experimental
information on the effect of raising the tensile-biased mean stress
(and reducing the amplitude), although the fracture mechanics model
predicts that the endurance is increased at a particular peak stress
level.

This model also predicts that when crack growth rates are similar, higher strength materials will show less endurance than weaker ones, in conflict with Wilkins' data. Wilkins and Jones [103] discuss this point in terms of the interaction between growing cracks in the real material rather than the growth of a single flaw in an ideal sample. Wilkins' statistical approach, with its different definition of homologous stress, σ_h, considerably reduces the inherent scatter in the experimental data and provides the engineer with a realistic method of estimating the cumulative probability of fatigue failure.

Both Wilkins' approach and Marshall and Priddle's fracture mechanics approach should be applied to extend data to other mean stress levels to further both the definition of design data and a mechanistic understanding of the fatigue process at the macro level. The engineer needs a detailed knowledge of how the fatigue characteristics vary with mean stress level and stress amplitude and how to sum the effects of varying stress conditions. It is the growth of macrocracks which are of concern to the design engineer.

A further important point is that many practical cases involve strain cycles, and these have not received experimental investigation. Strain cycles are likely to be less damaging than corresponding stress cycles due to stress relaxation in the early stages, but should receive experimental verification. The effects of temperature appear to be beneficial, but data are very limited.

VIII.   NOTCH SENSITIVITY

Failures of components in service are often associated with geometrical stress-raising features, and it is necessary in design to evaluate their effect. In the absence of adequate theoretical treatment, empirical methods are often employed by determining experimentally the notch sensitivity of the material under tensile stress. The perturbation of the stress field by the notch is really an extreme example of the influence of stress gradients and for graphite, an analysis based on a maximum stress criterion of failure would almost

certainly give pessimistic results for the component failure conditions. We have already seen that graphite specimens under tension
or bend are insensitive to the presence of narrow-edged slits of
depth up to a critical value determined by structural features of
the material, apparently strongly related to the grain size.

The theoretical stress-raising effect of a notch depends not
only on the radius at the root of the notch, but also on other geometrical features. The effect of different specimen and notch geometries on notch sensitivity has been examined for the near-isotropic
graphite grade IM1-24 in tension and bend, and since there are little
available data on this topic, the results are described here in some
detail.

Figure 20 shows the types of specimen used. Tensile specimens
varied from large double-edged notched plates of rectangular section
(unnotched breadth D, depth W) to circumferentially notched round
bars (maximum diameter D, minimum diameter d at the notch root).
Both 4- and 3-point bend tests were performed on single edge-notched
rectangular beams, the latter on a fine-grained version of the same
graphite. Some tensile specimens were tested after a low-temperature,
fast-neutron irradiation, which drastically reduced the plastic component of the deformation, caused the stress-strain curve to become
more linear initially and raised the strength by a factor of two
(Fig. 14).

Table 6 gives specimen details and the nominal stresses at
failure, which for tensile tests is the mean stress in the notch
section and for bend tests is the maximum stress assuming a bend
specimen of depth equal to the ligament length below the notch.
These nominal failure stresses are compared with the tensile or
bend strength (as appropriate) determined on unnotched specimens,
and the experimentally observed stress concentration factor is
defined as:

$$k_{(observed)} = \frac{\sigma_{(tensile \ or \ bend)}}{\sigma_{(nominal)}}$$

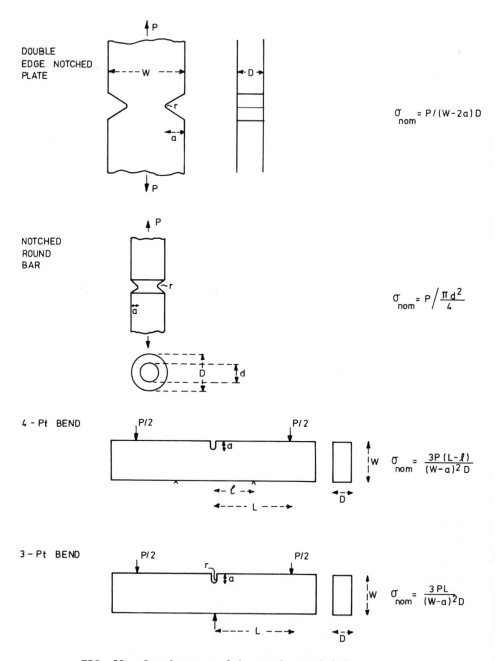

DOUBLE
EDGE NOTCHED
PLATE

$$\sigma_{nom} = P/(W-2a)D$$

NOTCHED
ROUND
BAR

$$\sigma_{nom} = P\left/\frac{\pi d^2}{4}\right.$$

4 - Pt BEND

$$\sigma_{nom} = \frac{3P(L-\ell)}{(W-a)^2 D}$$

3 - Pt BEND

$$\sigma_{nom} = \frac{3PL}{(W-a)^2 D}$$

FIG. 20.   Specimens used in notch sensitivity tests.

TABLE 6

Notch Sensitivity Data for IMI-24 Graphite

Block No.	Test mode	Specimen description[c]	Maximum diameter	Breadth D	Depth W	Notch Depth a	Notch Radius r		$\sigma_{(bend)}$ or $\sigma_{(tens)}$ (MN/m²)	SD	$\sigma_{(nom)}$ (MN/m²)	SD	$k_{(obs)} = \dfrac{\sigma_{(bend/tens)}}{\sigma_{(nom)}}$	k_{th}	$n = \dfrac{k_{obs}}{k_{th}}$
1	Tensile	Round bar, unnotched	10					6	23.9	2.1					
		Plates, double edge-notched[a]		25	50	12.5	10	4			17.8		1.34	1.5	0.90
		Plates, double edge-notched[a]		25	50	12.5	2.5	5			13.3		1.80	2.3	0.79
		Plates, double edge-notched[a]		25	50	12.5	0.6	5			11.9		2.00	3.7	0.54
		Plates, double edge-notched[a]		25	50	12.5	2.5	5			13.0		1.84	2.6	0.71
		Plates, double edge-notched[a]		25	32	12.5	2.5	7			20.6	1.7	1.16	1.4	0.83
		Round bar, circumference notched	9			2.0	10	7			24.8	2.3	0.96	1.1	0.88
		Round bar, circumference notched	9			2.0	2.5	7			22.5	2.3	1.06	1.4	0.76
		Round bar, circumference notched	9			2.0	0.6	6			19.1	1.9	1.25	2.1	0.60
2		Round bar, unnotched	10					6	23.1	1.5					
		Round bar, circumference notched	14			2.5	2.5	4			22.5		1.03	1.6	0.65
		Round bar, circumference notched	14			2.5	0.6	4			16.0		1.44	2.9	0.50
2 (irrad)		Round bar, unnotched	10					5	47.2						
		Round bar, circumference notched	14			2.5	2.5	3			38.6		1.22	1.6	0.76
		Round bar, circumference notched	14			2.5	0.6	3			33.5		1.41	2.9	0.49
2	4-point bend L = 19 mm ℓ = 9.5 mm	Beams, unnotched		12.5	12.5			7	36.1	1.8	35.6	0.9	1.01	1.2	0.84
		Beams, edge-notched		12.5	12.5	2.5	10	6			28.6	1.8	1.26	1.8	0.70
		Beams, edge-notched		12.5	12.5	2.5	2.5	6			24.8	0.6	1.45	3.1	0.47
		Beams, edge-notched		12.5	12.5	5.0	0.6	6			33.2	1.4	1.09	1.7	0.64
		Beams, edge-notched		12.5	12.5	5.0	2.5	6			26.9	0.9	1.34	2.9	0.46
		Beams, unnotched		6.2	12.5			8	37.2	1.7	30.3	1.4	1.23	1.9	0.65
		Beams, edge-notched		6.2	12.5	1.2	2.5	6			30.9	1.1	1.20	1.8	0.67
		Beams, edge-notched		6.2	12.5	3.7	2.5	6			32.7	1.6	1.14	1.75	0.65
		Beams, edge-notched		6.2	12.5	5.0	2.5	6			36.2	1.9	1.03	1.7	0.61
3	3-point bend	Beams, unnotched		12.5	12.5			13	33.5	1.7	33.6	1.1	1.00	1.3	0.77
		Beams, edge-notched		12.5	12.5	2.5	10	8			28.0	0.6	1.20	1.9	0.63
		Beams, edge-notched		12.5	12.5	2.5	2.5	8			22.2	1.0	1.51	3.1	0.49
		Beams, edge-notched		12.5	12.5	2.5	0.6	8			21.8	1.2	1.54	(6.2)	(0.25)
		Beams, edge-notched[b]		12.5	12.5	2.5	~0.02	8			20.2	0.7	1.66		
		Beams, edge-notched[b]		12.5	12.5	0.6	~0.02	8			22.1	0.3	1.52		

[a] 90°V notch.

[b] Sharp V notch.

[c] All others U shape.

This ratio is evaluated in Table 6 from the mean values from each group of specimens.

The large plate specimens were more notch sensitive than the small round bars, and the effect of a given notch was reduced by decreasing the notch separation, but the effect of notch shape was not marked. For all types of specimen, changes in notch root radii had a significant effect on the observed stress concentration factor, which had a maximum value of 2.0. The notched tensile specimens included those irradiated by fast neutrons with very different tensile strength but showing similar experimental stress concentration factors for both the root radii tested.

Table 6 compares the observed stress concentration factors k_0 with theoretical values k_t for an elastic material derived from the literature [104, 105, 106] for each geometry. For the large plate and some of the small round tensile specimens, k_t was confirmed by calculations using a finite-element program. k_0 is always less than k_t and never exceeds 2.0 for the material tested. Thus under the stress gradient near the notch root, the theoretically calculated maximum stress at failure exceeds the strength of the unnotched specimens and further illustrates the stress gradient effects discussed in Sec. V.

It is not possible in a general stress problem to specify the stress-raising effect of a given notch radius, since the stress concentration factor is also dependent on other geometrical factors. The best way of rationalizing the data is to compare the normal tensile (or bend) strength of the material with the theoretical value of the maximum stress at failure by evaluating their ratio, defined here as notch sensitivity n. This parameter is also the ratio of the observed stress concentration factor to the theoretical value, i.e.,

$$\text{Notch sensitivity } n = \frac{\sigma_{\text{(tensile or bend)}}}{\sigma_{\text{(maximum)}}} = \frac{k_0}{k_t}$$

Figure 21 shows that n may reasonably be expressed as a function of notch radius for all the different types of specimen examined. At high values of radius r, where there is very little notch effect

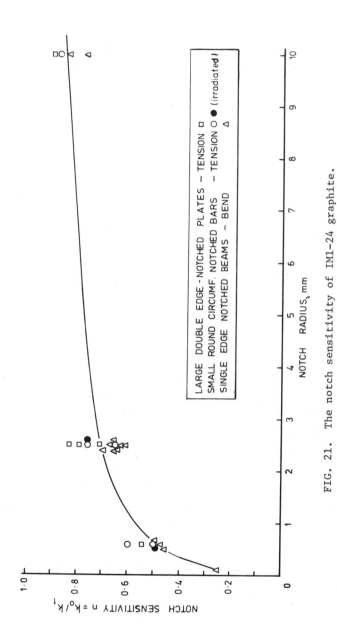

FIG. 21. The notch sensitivity of IM1-24 graphite.

($k_t \rightarrow 1$), the value of $n \rightarrow 1$ and the maximum stress approaches the tensile (or bend) strength. However at small values of r, although k_t may be high (a maximum of about 3 in present examples), the full stress-raising effect is not present, and n is small. In the limit, $k \rightarrow \infty$ and $n \rightarrow 0$. The observed stress concentration factor is nk_t, and as a practical upper limit, it is worth noting that the maximum observed stress concentration was a factor 2.0 in these tests, an upper limit on the effective sharpness of the notch being set by the microstructure of the material.

Tucker [107] has developed a theoretical model of a double-notched plate, using a fracture mechanics approach to investigate the effect of specimen geometry on the notch sensitivity of brittle porous materials. His analysis produces a relation between notch sensitivity (as defined here) and notch radius in good agreement with the experimental data in Fig. 21, assuming an inherent defect size of ≈ 1.4 mm, which is not unreasonable for these materials which have a grain size of ~ 1 mm.

These experiments only cover one graphite type, and Fig. 21 is not expected to represent a unique curve for all graphites but to vary with the structural characteristics of the material. Intuitively, it is expected that graphites with a coarse-grained structure would be least sensitive to small notches. Bazaj and Cox [108] report an increase in the observed stress concentration effect with decrease in the grain size of the material associated with an increased strength, and this experimental observation would also be predicted by Tucker's theoretical model. They also observed experimental stress concentration factors less than theoretical values which were substantially independent of temperature up to 2000°C for the fine-grained GLC grade H205 graphite (<0.75 mm) except for the smallest notch-root radius at this maximum temperature, where the experimental value is lower than those at room temperature and 1000°C (≈ 1.3 compared with 1.5). Their data for this graphite expressed in terms of n are in reasonable agreement with Fig. 21.

IX. EFFECT OF DENSITY, GRAIN SIZE,
AND CRYSTALLINITY

We have seen that ultimate macrofailure in graphite is preceded by
the development of microcracks. If an applied external uniform
stress is increased, the density of microcracks increases to a critical value when coincidental alignment gives an effective large
Griffith crack and complete fracture takes place. We have also
discussed the evidence for a statistical description of fracture
governed by the distribution of flaws and considered the implications
of this description when nonuniform stress fields are applied. Under
such nonuniform stress conditions, the density of microcracking will
vary in proportion to the local stress, and this will limit the ability of macrocracks to grow by coincidental alignment. The damaged
region at the tip of the macrocrack also explains the high absorption
of energy in crack growth and apparent lack of notch sensitivity.

The crack theory of fracture and the formal relationship given
by the Griffith fracture model serves as a basis for discussing the
effects of different variables and external influences on the strength
of graphite, i.e., $\sigma^2 c \sim \gamma E$, where σ is the tensile strength, E is
Young's modulus, c is the effective length of the critical crack,
and γ is the surface energy required to create new fracture surface.
These effects are most sensibly considered by examining the corresponding changes in Young's modulus, although in some experiments
this information is not available.

It must be remembered, of course, that Young's modulus samples
the whole structure, and hence modulus changes depend on a change in
average properties while strength changes depend on the change in
maximum flaw size. This implies that it is possible to obtain an
increase in strength for little change in modulus due to the elimination of the maximum defects without changing the main structural
characteristics of the material. Hence samples of constant density
can have similar moduli but different strengths due to variation in
the maximum crack length c.

The relationship between tensile strength and modulus is certainly not unique for all graphites, since differences in basic grain structure, density, degree of crystallinity, etc., result independently in different values for the parameters in this Griffith relationship. For the same basic structure, however, it is informative to examine the effect of independent variables, and even to try to assess the relative contribution of combined effects imposed on the material.

Hutcheon and Price [109] found that successive impregnation of a petroleum coke graphite raised the bend strength by a factor of about three and Young's modulus by a factor of about two. The relationship between strength and total voidage was of the form $\sigma_b \propto \varepsilon^{-1.6}$.

Losty and Orchard [11] observed that on successive impregnation of a reactor grade graphite, the bend strength and Young's modulus increased such that $\sigma_b \propto E$, and the strain to failure remained constant. Thus the strain energy at failure increased with increasing density. They interpreted these results as consistent with a constant strain energy at failure within the original graphite grains, with the impregnant acting as an additional body in parallel, i.e., a constant strain model with no change in the modulus of the grain. In his discussion of this work in terms of the Griffith equation, Mason [50] pointed out that impregnation would increase the average density and reduce the crack length. Both of these effects would cause an increase in strength, but only the average density change would affect Young's modulus. The energy density per unit volume σ^2/E necessary to produce a given critical energy density in a localized region will rise in proportion to density and inversely in proportion to crack length. Mason gave a qualitative picture of the interrelationship between strength and Young's modulus, pointing out that density and crack length could vary independently and the correlation between the two properties would differ accordingly. For example, test pieces of constant density, but different maximum flaw sizes will show a scatter in strength but a similar Young's modulus.

Losty and Orchard [11] also demonstrated by two independent experiments that when changes were induced within the graphite crystallites without affecting the crack structure external to them, then Young's modulus and bend strength varied in such a way that the elastic strain energy to failure σ^2/E remained constant, in accordance with the Griffith equation for no change in (γ/c). In heat treatment experiments on carbon, the variation in degree of graphitization in samples of constant density, which affected the crystal properties but not the grain structure, caused Young's modulus to decrease by a factor of about two and the bend strength to vary in such a way that σ^2/E remained constant. The same relationship between strength and modulus applied in reactor grade graphite samples which were irradiated by fast neutrons under low-dose, low-temperature conditions, which caused a modulus increase (by a factor of two) due to the pinning of basal plane dislocations within the graphite crystallites, but no change occurred in the grain structure. The same proportionality factor between σ^2 and E was found in both parallel and perpendicular directions with respect to extrusion for this anisotropic material. These experiments support the view that failure occurs at a constant strain energy within the graphite grains. High-dose, fast-neutron irradiation produces structural changes superimposed on the within-crystal effects, and this more complex behavior is discussed later.

Having established the effects of heat treatment and of impregnation separately, Losty and Orchard went on to show that the combined effect of these two processes was as if each acted independently. Their work led them to propose a model for the deformation of graphite to the point of failure consistent with their strength studies and their earlier work on elastic behavior. They argued that an applied stress is accommodated in the polycrystalline material by elastic shear deformation within the grains, accompanied by slip. They believed that the accumulation of slip is not the ultimate cause of failure, but that failure occurs when sufficient strain energy is established within the grain to perform the work necessary to propagate a crack across the misoriented grain boundaries.

When material is removed rather than added to alter the macro-structure, a similar relationship between strength and modulus is observed. An almost constant strain to failure is obtained when graphite is oxidized radiolytically in CO_2 so that the oxidation attack is substantially uniform. This effect was observed in British Pile Grade A graphite by Hawkins [110] and in near-isotropic graph-ites by Brocklehurst et al. [111], and the *relative* changes in strength and modulus with weight loss were the same for the differ-ent types of graphite examined and independent of the mode of test-ing. Attempts to describe the modulus change on existing models for porous bodies show that the modulus falls more rapidly than can be accounted for by an increase in size of isolated voids, and implies that a substantial fraction of the voidage is continuous throughout the matrix.

Under thermal oxidation, the changes in strength and Young's modulus also indicate failure at constant strain. Board and Squires [112] presented tensile, bend, and compressive strength, with Young's modulus data for British Pile Grade A graphite oxidized in CO_2, and concluded that failure occurs at constant strain. Rounthwaite et al. [113] also showed strength data for the same material oxidized in CO_2. They made direct measurements of tensile strain to failure and also concluded that there is very little change in this parameter with thermal corrosion. However, the changes in both strength and Young's modulus for a given weight loss by thermal oxidation are much greater than corresponding changes under radiolytic oxidation. Figure 22 compares data obtained under both types of attack. Under thermal corrosion, both these papers reported that the binder was preferentially oxidized, possibly due to a larger specific surface area, and observations by Board and Squires of crack propagation in bend specimens, using Jenkins technique discussed earlier, showed that while fracture occurred by linking of pores by cracks through the large grains in unoxidized material, the cracks tended to prop-agate entirely through the binder or along grist-binder boundaries in highly oxidized material.

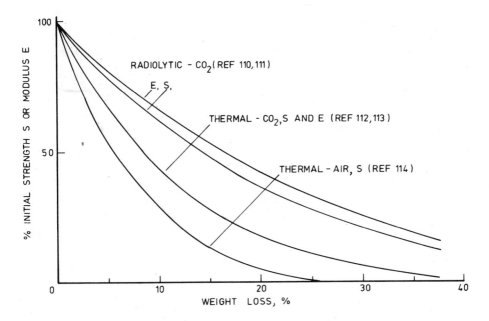

FIG. 22. Comparison of the effect of thermal and radiolytic oxidation in CO_2 and of thermal oxidation in air on the mean strength and modulus of graphite.

The effect of thermal oxidation in air was studied for PGA graphite by Knibbs and Morris [114], who found a more severe deterioration in bend strength at a given weight loss for oxidations performed at 470°C. Their observed strength-weight loss relationship is compared with those obtained in CO_2 in Fig. 22. They attributed this effect to preferential attack possibly at grist-binder interfaces, and the experiments indicated a gradual weakening of the binder-grist coherence, with surface erosion which may have influenced the bend strength results.

Armstrong [115] has reported the relationship between strength and Young's modulus for a series of near-isotropic graphites prepared from Santa Maria coke filler with porosity as the major variable. The strength data showed considerable scatter, and he rightly argued that the modulus was insensitive to the large flaws producing low strengths, and defined the maximum strength bound to the data obtaining the following relationships between Young's modulus E, and (1) compressive strength σ_c and (2) flexural strength σ_b:

$$1. \quad \sigma_c \times 10^{-3} = -5.39 + 14.7 \text{ E} \times 10^{-6})$$
$$2. \quad \sigma_b \times 10^{-3} = -2.31 + 6.29 \text{ E} \times 10^{-6}) \quad \text{units lb/in.}^2$$

The ratio between the two strength values was constant, but the strain to failure tended to increase for the higher strength and modulus materials. It should be noted, however, that all the other experiments showing failure at constant strain referred to density variations caused by impregnation or oxidation of the same basic structure.

Knibbs [31] used literature data to examine the relationship between bend strength and grain size for 12 different graphites varying in maximum grain size by two orders of magnitude. He allowed for density difference by assuming an exponential law, and fitted the data to the empirical equation.

$$S = k \, d^{-a} \, e^{-bP}$$

where

S = average strength (dyn/cm^2)
P = fractional porosity
d = maximum grain size (cm)

He found the constants to be $k = 1.2$, $a = 0.5$, and $b = 6.8$. Drawing attention to the square root dependence on grain size, he suggested that the Griffith failure criterion could be applied to polycrystalline graphite with the critical crack length identified with the maximum grain size.

In contrast to the relationship observed by Knibbs for materials with maximum grain sizes ≤1 mm, Amsez and Volta [116] showed that for a range of industrial graphites the reduction in 4-point bend strength with increase in maximum grain size from 0.4 to 6.7 mm was only a factor of about two and not four as expected from the grain size difference. However, since all measurements were performed on specimens cut parallel to the extrusion axis, it is possible that changes in anisotropy by an increased degree of particle alignment

might have played a strengthening role in the very coarse-grained
material. They also observed that the dependence of bend strength
on density was a function of grain size, being most marked for the
finer-grained materials, but again it was not clear how to allow
for any differences in anisotropy. For bend specimens notched to a
depth >10% of the specimen thickness (notch depth >3.2 mm), the
strength was independent of grain size, and the influence of density,
relatively small. Their measurements of the effective work of fracture by analytical and compliance techniques showed that G_c increased with increasing density but appeared to be independent of
grain size. (Values obtained by the former technique varied from
about 80 J/m^2 at 1.6 g/cm^3 to 150 J/m^2 at 1.8 g/cm^3, while corresponding values from the compliance technique were considerably
higher, about 120 and 250 J/m^2, respectively.)

Thus most of the data indicates that changes induced within
the macroporosity external to the basic crystal structure cause
failure at constant strain, which can also be interpreted as either
a constant strain or a constant strain energy at failure within the
grains; there are, however, apparently some exceptions to this general rule. Strength increases with decreasing grain size, and for
many materials a square root dependence is observed, but the rate of
change in strength with density is decreased for large grain sizes,
presumably as a result of a decreasing surface to volume ratio of
the grains.

X. EFFECT OF TEMPERATURE, ATMOSPHERE, AND STRAIN RATE

A. *Temperature and Atmosphere*

It is well known that the strength of graphite increases with temperature showing a maximum at about 2500°C at which stage creep
effects become predominant. Malmstrom et al. [117] and Martens et
al. [118] were among the early workers demonstrating this effect.
Values from this early work are cited by Nightingale [119] showing
maximum increases at 2500°C by factors of 1.6 to 2.8 for different

grades of graphite, while the elastic breaking strain of one grade
was reported constant over the range 0 to 2000°C. Various explanations have been advanced for these observations. Mrozowski [6]
postulated that the large anisotropy in thermal expansion coefficient
of the graphite crystallites would cause the generation of cracks
and/or frozen-in stresses on cooling from the graphitizing temperature. The observed increase in strength on heating is then attributed to the relief of these stresses arising from anisotropic contraction. Martens et al. [120] attributed the strength increase
to a continuous increase in plasticity of the graphite reducing the
stress concentrations at pores or other defects in the structure.

Microcracks parallel to the crystallite basal planes, i.e.,
suitably oriented to be formed by the large c-axis contraction, have
been observed in electron microscope studies. This oriented microporosity is seen to close on fast-neutron irradiation due to crystallite c-axis growth [121], and successful theories relating crystal
and polycrystalline dimensional change behavior [122] are based on
an interpretation of the important structural influence of such
microcracks.

The thermal closure of these microcracks has formed the basis
of explanations for the increase in Young's modulus of graphite with
temperature [123]. This increase occurs, even though the crystal
moduli all decrease with temperature. Andrew and Sato [124] showed
that the modulus variation with temperature is strongly dependent
on the nature of the filler coke but not on the binder material.
"Hard" filler particles which do not readily graphitize showed little
change in modulus up to 1300 to 1400°C followed by a continuous decrease in modulus with further increase in temperature. However,
"soft" filler materials which easily graphitize showed a rise in modulus with increasing temperature and a maximum below the heat treatment temperature (just below 2000°C in the specimens graphitized at
3000°C). The maximum represented an increase of 40 to 60% over the
room temperature value, and the increase was about 17% at 1000°C.
Davidson et al. [125] show for well-graphitized materials based on
petroleum coke and pitch coke an increase of about 10% at 1000°C

rising to about 40% near 1900°C, but smaller increases were observed
on other materials.

The influence of the oriented microporosity on the deformation
characteristics is reviewed by Jenkins [1], although he omitted the
contribution made by Mason and Knibbs [123], who described their
studies of the temperature dependence of elastic modulus of carbons
and graphites from -196°C to 1000°C based on a model in which the
elastic properties were dominated by the crystal shear modulus c_{44}.
The results showed a minimum in the temperature dependence at about
300°C, and the subsequent rise in modulus at the higher temperatures
was attributed to a crack-closing process determined by the difference in volume thermal expansion coefficient of the polycrystalline
material and the constituent crystals. However, in considering the
strength data in detail, the effect of test atmosphere will be considered first, since it has been shown to play an important role in
strength determinations and to influence the interpretation of the
temperature dependence at moderate temperatures.

Diefendorf [126] showed that in bend tests on a commercial
petroleum coke graphite (Speer 580), the strength at 1000°C in vacuum
was about 20% greater than at room temperature in air. On cooling
to room temperature while maintaining the vacuum, about half this
strength increase was retained, leaving a residual increase of about
10%, but the readmission of air reduced the strength to the original
value. Rowe [127] also showed that after degassing at high temperature both modulus and strength were increased on testing at room
temperature under vacuum, but reduced again to the original values
when air was readmitted.

These effects of atmosphere, and their implications that gas
adsorption on accessible internal pore surfaces plays an important
role, suggest the possibility of strain rate effects on strength if
gas diffusion within the micropores on straining is involved.
Diefendorf [126] went on to demonstrate that increasing the strain
rate at room temperature by several orders of magnitude also caused
the strength to rise by about 10%, i.e., the fast fracture eliminated
the atmosphere effect. He thus concluded that two effects of roughly

equal magnitude were responsible for the apparent temperature dependence up to 1000°C, namely, an effect of atmosphere, suggested to
be a stress corrosion phenomenon, and a real temperature effect
possibly due to the relief of internal stresses. He examined the
effect of strain rate on strength over five orders of magnitude in
rate, but found most of the increase occurred for anvil speeds
increasing between 0.2 and 2 in./min.

Logsdail [128] identified the important adsorbate as water vapor.
Working with British Reactor Grade A graphite, he investigated the
effect on flexural strength of different gaseous environments. He
concluded that the room temperature strength and strain to failure
were reduced by the physical adsorption of water vapor, removal of
which increased these parameters by up to 40%. The removal was
easily achieved by degassing at room temperature in high vacuum or
by heating for a short time (approximate minutes) at 200°C in poor
vacuum. The effect of temperature on the strength of degassed
samples was not very significant up to about 1000°C, but then increased by 12 to 15% at 1250°C. A summary of these data is shown
in Fig. 23. In contrast to Diefendorf's results, there was no significant effect on room temperature strength when the strain rate
was increased over a similar range of anvil speeds. Logsdail suggested that this difference might be a consequence of the coarser-
grained structure of the graphite used in his work.

The effect of the water vapor was related by Logsdail to the
decrease in surface energy γ of the graphite caused by adsorption.
Using water adsorption data for a similar graphite he estimated the
decrease in γ in the Griffith equation to be about 0.07 J/m^2 at
room temperature. Taking a value for the surface energy of a graphite
crystal parallel to the basal planes of 0.15 J/m^2 [46, 47], the
calculated lowering of strength due to water adsorption was in good
agreement with the experimental values and is thus explained as an
influence on the weak Van-der-Waals bonding between the basal planes.
The energy of surfaces perpendicular to the basal planes is considerably larger than that of parallel surfaces [5.5 to 6.3 J/m^2 (46)],
and the calculated decrease of 0.07 J/m^2 is comparatively small so

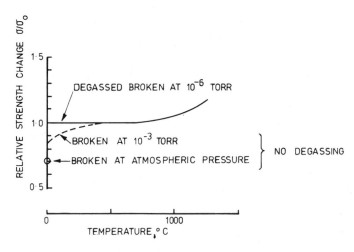

FIG. 23. Illustration of degassing effects on bend strength observed by Logsdail [128].

that the bond strength within the basal planes would be relatively unaffected. Experiments by Diefendorf [129] on pyrolytic graphite in different environments and orientations supported this argument, since they showed that only the strength between the basal planes was affected (reduced) by exposure to air, and the effect was also found to be rate sensitive with a high strain rate removing the atmosphere effect. (A typical strength at room temperature was ~160 MN/m^2.)

In their studies of the work of fracture of RC4 and AXF-Q1 graphites discussed earlier, Vitovec and Stachurski [93] reported that water soaking of the specimens produced only slightly lower values from the work of fracture measured in air at room temperature, but removal of the water by testing in vacuum raised the values by about 40% for RC4 and 15 to 30% for AXF-Q1. They attributed the effect to water adsorption in the pores and observed influences of anisotropy in the pore structure. Testing RC4 graphite at 750°C in vacuum produced a similar result to the room temperature value in vacuum. The temperature dependence of work of fracture measured by Udovskii et al. [86] up to 2300°C has been discussed earlier, showing increases by factors up to 10 (see Table 5).

The increase in fatigue limit with temperature reported by
Green [97] and discussed in the appropriate section may well be
associated, at least in part, with an atmosphere effect. The influence of atmosphere on fatigue behavior has not been investigated.

Gillin [62] studied the temperature dependence of the compressive strength of both an isotropic graphite and Grade A nuclear
graphite. There was little change up to 1000°C then a rise in
strength with a maximum at about 2100°C, showing an increase of about
50% for the former and about 30 to 50% for the latter material, depending on direction. Fracture occurred in a plane inclined at 45°
to the load axis, and at room temperature the fracture strength of
both materials in air was about 5 to 10% lower than that in vacuum.

Tensile strength data were reported by Reynolds [130] for a
reactor graphite. The strength maximum was at 2500°C with a factor
of two increase in strength, most of which occurs above 1000°C.

Pears [131] has reported the temperature dependence for the
tensile properties of three graphites, AXF-5Q, ATJ-S, and G90. The
strength showed a steady rise, accelerating at high temperatures to
a maximum at about 2750°C, again by a factor of about two. The
strain to failure decreased slowly with increasing temperature and
then increased rapidly above 1700 to 1900°C while the Young's modulus
showed little change with temperature up to about 1900°C and then
decreased.

B.   Temperature and Strain Rate

Smith [132] has reported very detailed studies of the effect of
temperature and strain rate on the tensile strength of H4LM graphite.
Increasing the strain rate from 0.005 to 2.0/min tended to reduce
strengths at the higher temperatures, particularly above 2000°C.
Thus in helium, where the increase above room temperature was about
40% at 2000°C, the pronounced rise between 2000 and 2500°C at the
lower strain rate (giving an overall increase of a factor of nearly
three) was smoothed out at the higher strain rate to give a continuing steady rise and an overall increase of about 50%. Below 2000°C
smaller effects of strain rate were observed in the direction of a

weakening effect of higher strain rate at the higher temperatures,
mainly in the region 1000 to 2000°C, and there were detailed differences between tests in helium and in vacuum. However, the higher
strain rate removed these differences and gave a steady, almost
linear, strength rise between room temperature and 2500°C of about
50%. Smith [133] further showed that the effect of prestraining
under tension at 2500°C did not markedly affect the low strain rate
test results, but produced an increase in the high temperature-high
strain rate, tensile strength data. Simple annealing at 2500°C
without stress did not produce this strengthening. Figure 24 shows
a simplified representation of the major effects observed.

Smith's work led him to the following description of the mechanisms involved in high-temperature strength. Cooling of the graphite
from 2500°C generates internal stresses as postulated by Mrozowski
which cause a weakening. These stresses can be relieved by creep
processes at temperatures down to about 2000°C and strength values in
this region are very dependent on strain rate. Below 2000°C, creep
processes are slow and stresses are relieved by the formation of the
submicroscopic interlamellar cracks which increase in density as the

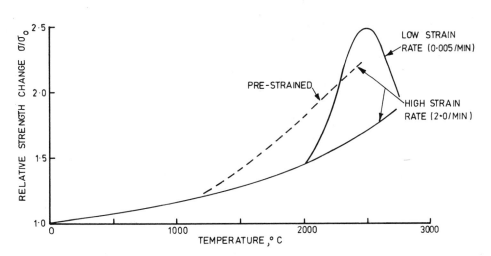

FIG. 24. Effect of strain rate on the relative increase in
tensile strength with temperature derived from data presented by
Smith [132, 133].

temperature falls. The rate of cooling may influence the temperature
for the onset of cracking and the crack density at a given temperature. Thus thermal annealing under stress above 2000°C can have a
strengthening effect, but below 2000°C can cause weakening. Subsequent reheating causes relief of internal stresses, and Smith argued
that it was not unreasonable to suppose that reclosing of the thermally produced microcracks between the crystallite basal planes
should result in a restoration of the Van-der-Waal's bonding, which
depends only on electrostatic forces. His experimental evidence
showed that the structural damage produced by cooling from 2500°C
was fully recovered by reheating. He explained the marked increase
in strength at low strain rates, which becomes large above 2000°C,
by a stress-directed self-diffusion of carbon atoms, which blunts
the subcritical cracks produced in the early stages of the fracture
process and observed under the optical microscope as described
earlier. It is clear that some form of stress concentration relief
is involved and at a different structural level. Gillin [62] attributed the high-temperature effect to the elimination of stress concentrations at tilt boundaries in the grist particles, which act
as barriers to dislocation motion.

The foregoing description is consistent with the observed temperature dependence of Young's modulus, which shows a maximum at
about 2000°C, i.e., the temperature at which the microcracking com-
mences on cooling. Other manifestations of the completion of crack
closure and onset of plastic deformation on heating to temperatures
about 2000°C are evident. For example, the sudden onset of very significant permanent sets at room temperature after thermal annealing
to successively higher temperatures is interpreted as a stretching effect on the structure by c-axis strain [134]. Also the sudden
rise above 2000°C in the rate of increase of the effective work of
fracture measured by Udovskii et al. [86], given in Table 5.

Smith [133] notes that the tensile fracture proces in H4LM at
temperatures up to 2000°C, and even up to 2750°C at high strain rates,
is identical to the earlier description of room temperature fracture
due to the growth of small, interlamellar, nonpropagating cracks

which increase in density until they ultimately join to form a fracture path. The ductility is low for this process and permanent
extensions to fracture are only a few tenths of a percent. At low
strain rates, however, the fracture mechanism changes above 2000°C
as tensile creep begins to operate and is characterized by the growth
of only a few cracks which extend directly to fracture. Permanent
extensions under such conditions can be a few tens of percent. The
reader interested in following the high-temperature creep behavior
of graphite is recommended to pursue the several papers produced
by the Los Alamos workers in this field, Green and co-workers [e.g.,
34, 35, 36, 135].

Having considered atmosphere and temperature effects with associated strain rate effects as necessary to the discussion, the latter
will now be considered in more detail.

Bazaj and Cox [108] give data for the strain rate dependence of
room temperature tensile strength for graphite grade H205 (Great
Lakes Carbon). The strength falls very rapidly at deflection rates
greater than about 5 in./min, estimated to be a strain rate of approximately 2/min, and in this region a doubling of the strain rate
halves the strength. The maximum strength was obtained in the region of 0.1 to 1/min, and at the other end of the scale, the strength
tended to fall at lower strain rates ~0.01/min.

Failure of graphite under very low strain rates and the extreme
condition of failure with time under constant stress (or static
fatigue referred to earlier) are possibly associated with the progressive diffusion of water to the tips of microcracks newly opened
up under the applied stress, reducing surface energy and causing
further crack propagation until a critical crack length for ultimate
failure is obtained. Whether this type of phenomenon occurs in fully
degassed graphite under vacuum has never been reported.

At the other extreme, high strain rates tend toward the conditions of impact and the common in-service design problems associated
with thermal shock, although these types of loading are generally
limited in imposed strain.

C. Impact and Thermal Shock

Very little work has been published on impact testing of graphite
because it is difficult to account for energy losses which occur by
processes other than the failure of the sample. The impact resistance of an elastic body is proportional to the elastic energy required to cause failure, and it has been reported [136] that the impact strength I ft-lb of a wide range of graphites is given by
$I = 4.63 \times 10^{-3} F^2/E$, where F is the flexural strength and E is
Young's modulus in lb/in^2 units.

Jones [137] has described and analyzed the results of a variety
of impact tests on graphite fuel element sleeves or on ring sections.
For repeated transverse impacts on ring specimens, an endurance curve
was obtained which was expressed in terms of the energy U of the im-
pacting hammer and a critical value U_0 below which failure did not
occur. After n impacts, the relationship was given by

$$n (U - U_0) = kU_0$$

with the constant $k \approx 2$. The critical energy U_0 for crack initiation
was approached asymptotically when $n \sim 100$ impacts and was related
to the elastic tensile energy at fracture in a static test. The
strength in one impact was related to the total work of fracture,
i.e., the total area under the stress-strain curve (compare unirradiated behavior in Fig. 14, assumed taken to failure). Hence in one
impact for this material, the ring absorbed up to three times the
elastic energy at fracture in a static test, and Jones attributed
this behavior at least in part to the nonlinearity in the stress-
strain curve and the energy absorbed in the deformation process.
Another contribution was proposed to arise from a stress gradient
effect which limits crack propagation during any one impact. More
precisely, the multiple impact strength U_0 is proportional to the
elastic energy per unit mass of material, i.e.,

$$U_0 \propto \frac{\sigma^2}{E\rho}$$

where

σ = static strength
E = Young's modulus
ρ = bulk density

Tests by Jones on thermally oxidized sleeves (up to 20% weight loss) gave fractional changes in impact resistance which agreed with predictions based on this relationship. In the earlier discussion, it was observed that for thermal and radiolytic oxidation failure occurred approximately at constant strain and $\sigma/\sigma_0 \approx E/E_0$, hence the fractional change in multiple impact resistance simplified to $U/U_0 = (\sigma/\sigma_0)(\rho_0/\rho)$.

Jones also tested fast neutron-irradiated sleeves after a peak fast neutron dose of 10^{21} n/cm^2 DNE at 300 to 400°C. The static strength and Young's modulus were both increased, but there was very little change in the elastic energy to static failure and hence in the critical energy in the impact tests, as predicted by the foregoing relationship. However, as the plastic component of the deformation was drastically reduced, there was a marked decrease in both the total work of static fracture and the first impact failure energy, which was only 1.7 U_0 instead of 3 U_0 in unirradiated tests. In contrast, the increased impact resistance observed by Sato and Miyazano [138] in lower-temperature irradiations, discussed with other irradiation data later, was associated with a similar increase in the elastic energy to failure.

The thermal shock performance of graphite is often evaluated in a comparative sense by experimental tests designed to rank graphites in order of merit, or by use of the "thermal shock parameter" $Sk/\alpha E$, where good performance is expected for a material with a high strength S, high thermal conductivity k, low thermal expansion coefficient α, and low Young's modulus E. The design evaluation of graphite performance under thermal shock conditions is important in many applications, and in such conditions when a limited strain is imposed, the effective surface energy of the material is probably more appropriate than the strength in determining the relative merits of different graphites. We have seen that some high-strength graphites tend to be more brittle, i.e., have lower surface energies, than some of the weaker materials for which a larger proportion of energy is dissipated by crack termination at pores and secondary cracking.

There is a parallel here in current views on thermal shock fracture in brittle ceramics. Hasselman's [139] approach to the thermal stress resistance of these materials is that when fracture initiation cannot be avoided, the extent of subsequent crack propagation must be minimized. He distinguished between the kinetic propagation of initially small cracks when excess elastic energy is available, such as might occur if a high-strength material were selected and which can result in a catastrophic reduction in mechanical integrity, and the quasistatic propagation of longer cracks causing only gradual loss in strength. To promote the latter type of behavior and improve the thermal stress resistance of a single phase material, he proposed that the effective Griffith flaw size in the material should be increased, and there is experimental evidence that increasing the grain size of some materials is beneficial. The extent of crack propagation under these conditions is minimized by maximizing $Gk/\alpha E$, and the microstructure of graphite is generally favorable to this type of crack propagation with a high density of cracks having a relatively large effective length. The microscopic observations of crack growth in graphite with increasing static stress and the associated decreases in strength, discussed in detail earlier, show similarities to the crack propagation and strength reduction in brittle ceramics due to increasing degrees of thermal shock as discussed by Hasselman.

XI. EFFECT OF FAST-NEUTRON IRRADIATION

There is now a wealth of data on the effects of fast-neutron irradiation on the nondestructive properties of graphite, but only a relatively small amount of data on strength. Where it does exist,
strength data are often limited in a statistical sense due to irradiation space limitations and competition with other requirements.
It must be recognized that, in general, contributions to irradiation-
induced changes arise due both to within-crystal effects caused by
lattice defects and to structural effects external to the crystal
caused by crystallite volume and shape changes. These effects must
be separated for any interpretation and are discussed in detail
later. Again it is useful, where possible, to discuss strength
changes in conjunction with corresponding modulus changes.

The initial effect of fast-neutron irradiation is to increase
both mean strength and modulus, and has been attributed to dislocation pinning by small, irradiation-induced defects within the graphite
crystal. The increases are initially rapid and tend to saturate at
a level which decreases with increasing irradiation temperature, although low-temperature, low-dose irradiations can show maxima in both
strength and modulus [140].

Some of the early low-dose work [140, 141] showed very similar
changes in both strength and modulus, suggesting that the elastic
strain to failure was initially constant on fast-neutron irradiation.
However, in a series of careful experiments discussed earlier, Losty
and Orchard [11] showed that when changes were induced within the
graphite crystallites without affecting the macrostructure by low-
dose, low-temperature irradiations, then modulus and bend strength
of a reactor graphite varied such that the elastic strain energy
to failure σ^2/E remained constant, consistent with the Griffith
equation for no change in γ or c. Matthews [41] also supported this
relationship between mean bend strength and modulus in irradiation
experiments at 400°C on AXF-5Q graphite.

In contrast, however, other work has indicated failure under
almost constant strain conditions. Sato and Miyazono [138] give
compressive, bend, tensile, and impact strength data from irradiations
at 30 and 80°C to a total dose of 7×10^{20} n/cm^2 on a reactor graphite
and draw attention to the relationship between strength σ and Young's
modulus E, before and after irradiation, which they represent by
the expression $\sigma \propto E^n$. The proportionality constant may be determined before irradiation and values for n were reported to be 1.2
for bending, 1.0 for compressive splitting (tensile), 1.4 for compression, and 1.3 for impact tests. The exponent for the impact
data particularly appears to be strongly influenced by one low modulus point and, in fact, most of the impact data would appear to
fit n $\approx$ 1.0. Factorial increases in all the strength and impact
data were in the range 2 to 4, and most of the damage was thermally
annealable. Under different irradiation conditions, Jones [137]
observed no improvement in impact resistance, as discussed in Sec. X.C,
when the elastic energy to failure was little changed by irradiation.

Irradiation experiments [30] on three graphites at 150°C in
DIDO showed that while Young's modulus increased by a factor 2.0
for all three materials, the tensile strength increased by a factor
$\approx$1.5 on one graphite and by 2.0 on the others, suggesting that the
irradiation-strengthening factors are not unique for all materials,
but in addition to the dislocation pinning effects within the crystal,
which should give σ^2/E = constant, are influenced in some cases by
irradiation-induced structural changes. This work also showed that
changes in compressive strength were ultimately larger than those
induced in tensile strength, and as stress-strain curves became more
linear with less plastic deformation, the compressive/tensile strength
ratio increased toward the value of eight predicted by the Griffith
theory for a perfectly brittle solid.

Platonov et al. [142] showed that in irradiations over the temperature range of about 100 to 570°C, the compressive strength of
a reactor graphite increased, with saturation levels decreasing with
increasing temperature from a factor of about 1.9 at the lower

temperature to about 1.2 at the higher temperature. The relationship
with modulus change in the well-graphitized material was dependent on
irradiation temperature, corresponding approximately to a constant
value of σ/E at the lower irradiation temperature, but a lower strength
change for a given modulus change at the higher temperatures. Allowing for scatter in the data, the relationship at the higher temperature was nearer to a constant strain energy than constant strain.
In less well-graphitized material, irradiation-induced modulus
changes were lower and dependent on heat treatment temperature, but
strength changes were not very different, showing that the relationship between modulus and strength during irradiation is dependent
on crystallite size. Thus in these experiments, large crystallite
size and high-irradiation temperature favor the constant strain energy relationship σ^2/E = constant, with strength changes lower than
the modulus changes. Lower crystallite size and lower temperatures
would both lead to increased retention of irradiation damage and
larger crystallite dimensional changes. Under these conditions,
strength changes and Young's modulus changes become more comparable
possibly because of a contribution from structural changes between
crystallites.

Engle et al. [143] presented data for H451 graphite, which indicated tensile strength increases by factors of about 1.7 to 1.5 after
irradiations in the range 600 to 1350°C, with the lower changes at
the higher temperatures. The spread of irradiated data indicates a
strength increase in all specimens. Corresponding factorial increases
in modulus were higher in the approximate range 2.3 to 1.6, and the
mean strength/modulus relationship was intermediate between constant
strain energy and constant strain, supporting the former at the lower
temperature (600°C) and tending toward the latter at the higher temperature (1350°C, maximum shrinkage rate).

At the Carbon 72 Conference in Baden-Baden (1972), M. R. Everett
and F. Ridealgh reported modulus and strength data for a Gilsocarbon
graphite irradiated in the temperature range 900 to 1200°C at

fast-neutron doses up to 3×10^{21} n/cm^2 (DNE).* Their data at these
high irradiation temperatures showed an almost linear strength-
modulus relationship with little decrease in elastic strain to failure. The tensile stress-strain curve after irradiation at 1200°C
to 10^{21} n/cm^2 was essentially linear.

All these data show a tensile strength-modulus relationship during irradiation between upper and lower bounds corresponding to failure at constant strain and constant strain energy, respectively.
High irradiation temperatures ($\sim$1000°C) favor the latter, intermediate
temperatures ($\approx$500°C) the former, and at low temperatures ($\approx$100°C),
both limits have been observed with differences between material
types in the same experiment. The contribution made by irradiation-
induced structural changes is difficult to evaluate in these experiments, and the separation of within-crystal and structural effects
is discussed in detail later in this section. The degree of heat
treatment also influences the subsequent relationship between the
irradiation-induced changes, possibly reflecting the effect of
crystallite size on the irradiation-induced defect pattern at a
given temperature.

Statistical presentations of data may be made when sufficient
data exist on the same material irradiated under the same nominal
conditions of temperature and dose, although it must be recognized
that increased scatter can be caused by the variation in irradiation
conditions within a large batch of samples.

Figure 25 shows the change in distribution of 3-point bend
strength of IM1-24 graphite caused by irradiation in DFR (Dounreay
Fast Reactor) to a fast-neutron dose of 2.5×10^{21} n/cm^2 (DNE) at
900°C. This irradiation also produced a mean shrinkage (with

----

*The fast-neutron dose in an experimental position is expressed in
terms of the equivalent damage dose in a standard position in the
DIDO reactor at Harwell (producing the same displacement damage to
the graphite) and defined by the nickel-activation reaction ^{58}Ni
(n/p) ^{58}Co, with a cross-section of 107 mb. This graphite damage
dose is thus referred to as the DIDO Nickel Equivalent dose.

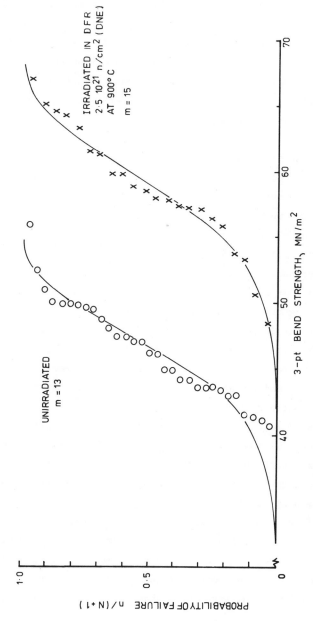

FIG. 25.   Effect of fast-neutron irradiation on the bend strength of IMI-24 graphite.

standard deviation) of (0.32 ± 0.04)%. The specimens were rectangular beams 36 x 9 x 4.5 mm with an outer support distance of 32 mm.
Approximately 50 specimens were machined and 22 selected at random
for irradiation, the remainder being used to define the unirradiated
strength distribution. Dynamic modulus measurements were also made
and Table 7 gives mean values of both strength and modulus with the
coefficient of variation (standard deviation/mean strength).

Figure 25 shows that the effect of irradiation is to shift the
bend strength distribution to higher values while maintaining a
similar shape. The standard deviation is 8.0% of the mean strength
for both unirradiated and irradiated specimens, while Weibull analyses of both the unirradiated and irradiated data (with $\sigma_u = 0$) give
Weibull modulus m values of 13 and 15, respectively, so that the
degree of homogeneity is virtually unchanged.

The fractional changes in Table 7 show that the mean strength
increase is almost exactly given by the square root of the modulus
change following the Griffith relationship for no change in the average size of the large, fracture-determining cracks. That is,
there is no significant increase in this effective critical crack
length due to structural changes, although the increased scatter in
Young's modulus, which samples all the cracks, suggests that the
overall crack pattern is changing.

At higher damage levels, however, structural changes are very
significant, and we now consider the relationship between Young's
modulus and strength changes for graphites subjected to very high,
fast-neutron damage levels. Figure 26 shows the fast neutron-induced
modulus changes for a wide range of near-isotropic potential reactor
graphites, of the types described by Nettley et al. [144], irradiated
in DFR to high doses in two temperature ranges at about 400°C, and
600 to 800°C. In order to understand these modulus changes, it is
necessary to consider the contributions which arise from within-
crystal effects and from structural changes external to the graphite
crystallites.

The initial rapid increase in Young's modulus has been attributed to the pinning of glissile basal plane dislocations by small

TABLE 7

Strength and Modulus of IM1-24 Graphite after Irradiation in DFR to 2.5×10^{21} n/cm^2 (DNE) at 900°C

Property	Unirradiated			Irradiated			Ratio, Irradiated/Unirradiated
	Mean (MN/m^2)	Coefficient of variation	No. tested	Mean (MN/m^2)	Coefficient of variant	No. tested	
3-point bend strength	46.4	8.0	31	58.7	8.0	22	1.27
Young's modulus	11.7	4.3	22	18.1	8.3	22	1.55

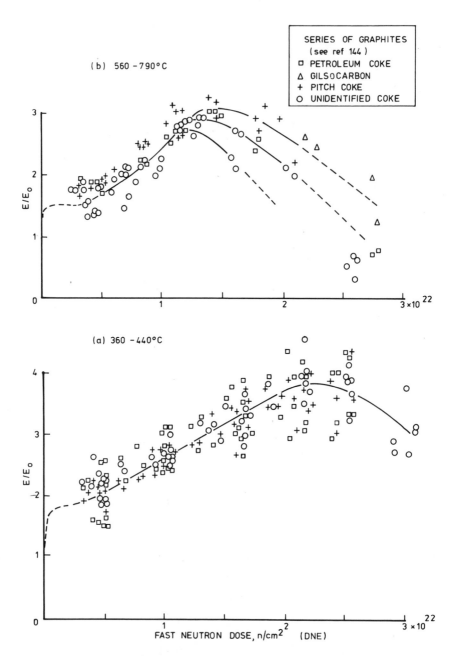

FIG. 26. Young's modulus changes with fast-neutron dose for different near-isotropic graphites irradiated in DFR. (a) 360 to 440°C; (b) 560 to 790°C.

irradiation-induced defects, which increase the effective shear modulus of the crystal. The equilibrium concentration of these small
defects is lower at the higher irradiation temperature and the
initial increase in Young's modulus is correspondingly lower. On
the dose scale in Fig. 26, this within-crystal effect is over very
rapidly but has long been established by several detailed studies in
lower-flux facilities and reported in various papers on graphite
irradiation damage.

Subsequent changes in modulus occur due to changes in the polycrystalline structure, illustrated in Fig. 27 by curves typifying
the irradiation-induced volume changes for these graphites. Initially,
there is a volume shrinkage, but the shrinkage rate decreases, and
reversal to growth occurs at high doses. As the irradiation temperature increases, the effect is accelerated, and shrinkage reversal
occurs at a lower fast neutron dose.

The initial shrinkage is a direct result of crystal dimensional
changes due to carbon atom displacements, discussed in detail else-
where [145]. At these irradiation temperatures the crystals change
shape rather than volume, the macroscopic effects are a balance
between the structurally determined contributions from c-axis growth
and a-axis shrinkage, and the initial effect on the structure is a

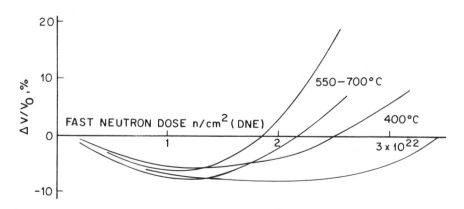

FIG. 27. Typical volume changes of near-isotropic graphites
irradiated in DFR.

closure of fine pores with the large pores relatively unaffected.
Young's modulus is sensitive to the closure of these small pores
during volume shrinkage and increases steadily as the porosity suitably disposed to accommodate crystal shear deformation decreases.
Other properties, for example, thermal transport, are relatively insensitive to the changes in these small pores. At higher doses,
unrestrained crystals will continue to change dimensions at constant
rates [146], but intercrystalline reactions in the polycrystalline
material result in the generation of large-scale porosity, the material shows a net growth, and the Young's modulus decreases. Thermal
conduction is sensitive to the development of these pores and also
decreases.

Thus closure of fine pores and opening of large pores occur
simultaneously with the effects of the former dominating at low doses
and of the latter at high doses, and the structural changes in volume
shrinkage are clearly not the reverse of those occurring on subsequent growth, but are observed as the resultant effect of two distinct
processes.

It is in fact possible to separate out the within-crystal and
structural effects on the total modulus change which is given by

$$\frac{E}{E_0} = \left(\frac{E}{E_0}\right)_{pinning} \left(\frac{E}{E_0}\right)_{structure} \tag{11}$$

In Fig. 26, which shows the total value, $(E/E_0)_{pinning}$ is approximately 1.8 at the lower temperature and 1.4 at the higher temperature
(levels defined more precisely in earlier work in lower-flux facilities. Hence $(E/E_0)_{structure}$ may be separated from the total change
as a function of dose, and this approach has been used in analyzing
irradiation creep data [147]. It is, however, considerably more
difficult to further divide the structural term to describe fine
pore-closure and large pore-opening effects separately.

Figure 28 shows relative strength changes as a function of fast-
neutron dose for a similar range of graphites. Tests were either a
tensile or a bend mode, and the points are individual results on
irradiated specimens relative to the unirradiated mean value for

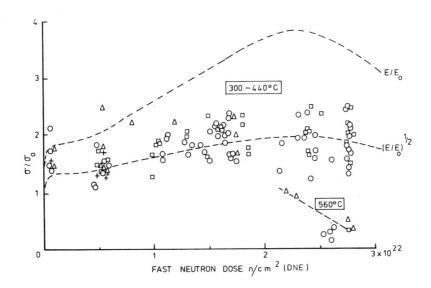

FIG. 28.   Tensile strength changes with fast-neutron dose for different near-isotropic graphites irradiated in DFR at 300 to 440°C and at 560°C. Key as Fig. 26.

the batch. The scatter in relative strength changes is higher than that in modulus and reflects both the uncertainty in the unirradiated value for the particular specimen tested, and the differences within and between the different materials. The unirradiated tensile strengths of these materials ranged from 10 to 22 MN/m^2, with typical standard deviations ranging from 10 to 20%, and there is an indication that the stronger materials show smaller initial factorial increases than the weaker materials.

The curves for the lower temperature data in Fig. 28 are given by either E/E_0 or $(E/E_0)^{1/2}$ from Fig. 26, and clearly the latter is the better overall fit at this irradiation temperature where the large-pore growth component is not very severe. At low doses, the square root relationship appears to define a lower limit to the mean strength data, but at high temperatures and doses, the fall in strength is greater than predicted by this type of relationship.

From the previous discussion on impregnation and oxidation effects, it would be expected that changes in the macrostructure might

result in a linear relationship between modulus and strength, if it
were possible to separate out the large-pore generation effect from
the irradiation data. Figure 29 illustrates that this linear relation probably does apply at very high irradiation damage levels when
large pore generation dominates. This figure shows the relationship
between modulus and tensile strength changes measured on the same
specimen. The arrows indicate the direction of increasing fast-
neutron damage. Initially, the square root relationship defines a
lower limit which persists as long as the increase in modulus continues. At low doses, the linear relationship is an upper limit
until strength changes reach a maximum factorial value of about
2.5. Clearly, the strength change does not continue linearly with
modulus over the full range of the modulus increase. At very high
doses, both modulus and strength decrease with dose as the graphite
grows, and the relationship is approximately linear as the large
pores develop.

It is interesting to note that at the lower irradiation temperature, where large pore-generation effects are not as great, none of
the individual strength data points in Fig. 28 lie below the

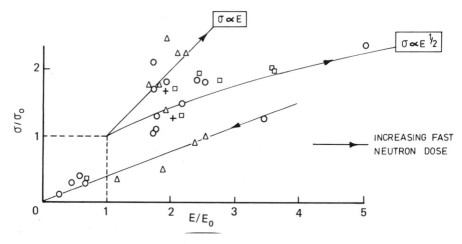

FIG. 29. The relation between tensile strength and modulus
changes with fast neutron dose. Key as Fig. 26.

unirradiated mean value, i.e., $\sigma/\sigma_0 > 1$ always. This observation
does not support the view that irradiation at low doses, i.e., below
shrinkage reversal, can cause an increased scatter and a reduction
in strength of some specimens, but it is clear that at higher doses
as dimensions revert toward the original values, larger strength reductions can occur.

The foregoing discussion has centered mainly on the relationship
between the changes induced in mean strength and in Young's modulus.
Design assessments of irradiation-damage effects must not rely entirely on mean strength behavior but must take account of the strength
distribution. In this light, data were presented earlier which
showed little change in the distribution of 3-point bend strength
under conditions where irradiation-induced structural changes were
not very significant.

Lungagnani and Krefeld [79] presented tensile data on a
Gilsocarbon graphite irradiated to various doses at 600 and 1150°C.
At the lower temperature, the mean strength increased continually
with dose within the dose range examined, but at 1150°C, the initial
increase was followed by a decrease to a level well below the unirradiated value, presumably as irradiation-induced structural changes
dominate. Their data at 1150°C showed that when the mean strength
had fallen again to a value comparable with the unirradiated value,
the risk of failure at low stresses was increased, i.e., the weaker
specimens were weakened more than the average by the irradiation-
induced structural effects. At this higher temperature particularly,
values of σ_u calculated from a Weibull fit decreased with increasing
dose, and with Weibull modulus m-values substantially constant at
3 to 5, indicated an increasingly greater risk of fracture at low
stresses for the irradiated material. Matthews [41] also reported
that in 400°C irradiations the Weibull predicted failure probability
at the 10^{-6} level for AXF-5Q graphite in bend tended to decrease,
even though the mean stress increased.

These data suggest that when structural changes become significant, the critical flaw distribution changes in a way such that their
mean size can remain relatively unaffected while the scatter in their

lengths is increased. Examination of the data points in these experiments does show, however, that in all cases where the mean strength
is substantially increased, i.e., before high-dose structural changes
become significant, no single experimental failure point is below the
minimum in the unirradiated distribution; this effect is only seen
experimentally when the bulk strength starts to fall again in
Lungagnani and Krefeld's experiments. The conclusions, therefore,
rely to some extent on Weibull fits to the data, with predictions of
the operating stress at low failure probabilities from relatively few
data points.

In using the Weibull analysis, the characterizing parameters σ_u
and m are often difficult to determine independently with any precision, and there is also a danger in extrapolating to probability
levels well beyond the range appropriate to the number of samples
tested. Thus conclusions drawn from such extrapolations should be
regarded with some care, but it is clear that when structural changes
are beginning to dominate and the mean strength is consequently falling due to the generation of large pores, there is a tendency for
increased scatter. Prior to this, however, when the mean strength is
still substantially higher than the unirradiated value, there is no
clear-cut evidence for a deterioration in the low-strength tail of
the distribution. Thus fast neutron-induced changes are complex,
compounded of within-crystal effects and structural changes external
to the crystallites. In nuclear reactor environments, additional
changes are superimposed due to radiolytic oxidation by the coolant.

Brocklehurst et al. [111] suggested that for design purposes
the cumulative effect of radiolytic oxidation and fast-neutron damage
on strength may be determined by assuming the processes are separable,
i.e.,

$$\frac{\sigma}{\sigma_0} = \left(\frac{\sigma}{\sigma_0}\right)_{irr} \left(\frac{\sigma}{\sigma_0}\right)_{ox} \tag{12}$$

This assumption could break down since the removal of material by
oxidation may affect the structural changes induced by irradiation,
possibly depending on the scale of the porosity attacked by the

oxidizing species, and in this case, simultaneous oxidation might produce a different result. However, it is of interest for design purposes to see how well Eq. (12) represents the cumulative strength change when the two processes operate in succession rather than simultaneously. Specimens of three types of graphite were pre-oxidized radiolytically in CO_2 in BR-2 and then irradiated in DFR in the temperature range 300 to 440°C. All specimens received a DFR dose of about 9×10^{21} n/cm^2 (DNE). From Fig. 26, it is seen that at this dose in the lower temperature range, the modulus change ~ 2.5. From the earlier discussion, corresponding strength changes due to irradiation alone, defined by lower and upper limits of $(E/E_0)^{\frac{1}{2}}$ and (E/E_0), respectively, should be in the range 1.6 to 2.5. The higher value also represents the maximum observed upper limit for tensile strength changes.

The effect of radiolytic oxidation alone on the strength of these graphites has been defined in Fig. 21, giving the term $(\sigma/\sigma_0)_{ox}$ in Eq. (12). Applying the above range of values for the factor $(\sigma/\sigma_0)_{irr}$ gives the range of expected values for the cumulative change. Figure 30 shows that the individual results obtained on irradiation of the pre-oxidized specimens (rings tested under diametral compression) are defined by the upper and lower predicted limits derived as previously.

At the Carbon 72 Conference at Baden-Baden (1972), R. Schill, G. Jouquet, and M. Yvars presented strength results on graphite which had been simultaneously irradiated and oxidized radiolytically. The graphite was based on Lima coke, and was irradiated in CO_2 at 400°C to various doses, suffering maximum weight losses over 60% of the original value, instead of a predicted reduction factor of 25% based on Eq. (12) and a strength change due to irradiation alone of $\sigma/\sigma_0 = (E/E_0)^{\frac{1}{2}}$. This predicted factor would only rise to about 35% had a relation $\sigma/\sigma_0 = E/E_0$ been adopted. Hence for this graphite, the foregoing relationships are very pessimistic in terms of predicting the strength reduction for a given weight loss, while giving a tolerable representation of experimental data obtained on to about 35% had a relation $\sigma/\sigma_0 = E/E_0$ been adopted. Hence for this

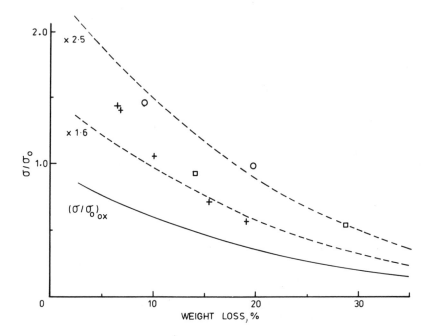

FIG. 30. Comparison of observed tensile strength changes in
oxidized-irradiated graphite specimens with predictions assuming
$\sigma/\sigma_0 = (\sigma/\sigma_0)_{ox} (\sigma/\sigma_0)_{irr}$. Key as Fig. 26.

other graphites. This difference suggests that the strength-weight
loss relationship is not unique for all materials; it may differ if,
for example, their pore spectra are widely dissimilar.

XII.   EFFECT OF INTERCALATION

Graphite will form interstitial lamellar compounds with several sub-
stances, and the reactions have been discussed, for example, by
Ubbelohde and Lewis [148]. These compounds retain the general graphite layer structure, with the planes of carbon atoms alternating in
a definite periodic sequence with atomic planes of the reactant.
Consequently, the intercalation is accompanied by an expansion of
the graphite crystal perpendicular to the layer planes. Strength
measurements on graphite specimens during the intercalation of
bromine have shown a marked decrease in strength with increasing

bromine content. However, to understand these results, a summary of
the general behavior of graphite under bromine intercalation is necessary.

The controlled intercalation of bromine by polycrystalline graphites, introducing a defined crystal strain, has been used [149, 150]
to study the influence of the polycrystalline structure on the relation between crystal and bulk dimensional changes for a wide range
of graphites. Comparison with other dilation processes, namely,
thermal expansion and fast neutron-induced changes showed that over
defined limits the polycrystalline dilation is determined by the
crystal strain and the same structural characteristics of the graphite in all these dilation processes. Most of the intercalated bromine may be readily removed by exposure to air, and there is substantial recovery of the dimensional changes with some hysteresis
effects. There remains, however, a residual bromine concentration,
which is difficult to remove and which profoundly alters certain
graphite properties such as the thermal expansion coefficient [151].
The correlation between bromine- and irradiation-induced dimensional
changes breaks down at high crystal strains, but in both dilation
processes the phenomenon of pore generation is observed in which the
bulk volume growth rate exceeds that of the crystals, generating new
porosity.

Figure 31 summarizes the linear dimensional changes on absorption
of bromine for British PGA graphite, giving the range for about 30
specimens cut in each of the directions parallel and perpendicular
to the extrusion axis of the material. Figure 31 also gives the
mean volume changes for this graphite and for a graphite crystal.
Figure 32 illustrates the extent of the deformation and recovery for
one parallel and one perpendicular PGA specimen (1) before test,
(2) at high bromine content, showing large dilation and distortion,
and (3) after substantial removal of bromine by exposure to air,
showing the high degree of strain recovery. Although not clear on
the photograph, the samples doped to high bromine content show surface cracks and crazing which disappear again on removal of the
bromine. This is one further example of the ability of the material

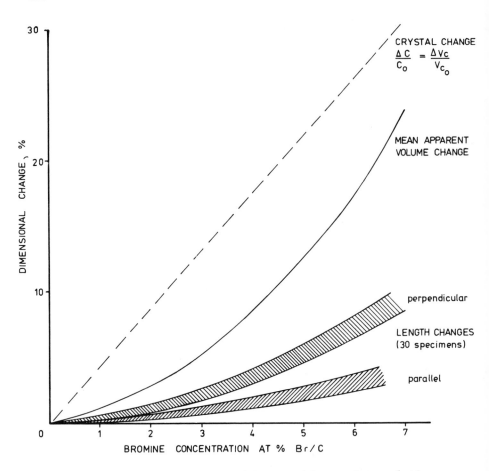

FIG. 31 Dimensional changes of PGA graphite on intercalation
of bromine.

to undergo extremely high strains and yet retain a memory of its
original state; more detailed bromine dilation and recovery curves
for this material and other graphites have been given [149, 150].

Figure 31 shows that the maximum apparent volume changes
$\Delta V_A/V_A$ are greater than 20%, and at these high strains, the rate of
change of apparent volume exceeds that of the crystal, hence generating pososity. The detailed studies showed that the crystal changes
are linear with bromine content at a rave $1/V_c \, dV_c/dB = 4.4\%$ at. %
Br/C and hence the net change in pore volume is given by

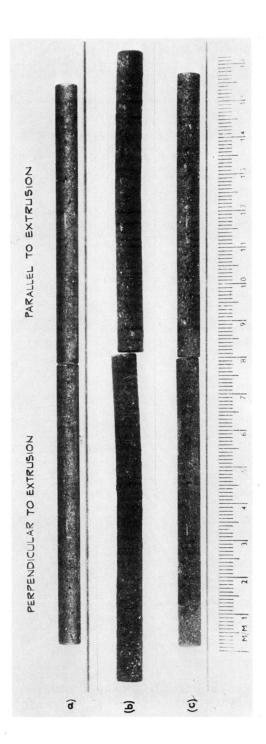

FIG. 32. Specimens of PGA graphite cut parallel and perpendicular to the extrusion direction (top) before exposure to bromine; (middle) after prolonged exposure to bromine vapor, showing increase in length and distortion; and (bottom) after removal of bromine showing recovery of strain.

$$\frac{\Delta V_p}{V_A} = \frac{\Delta V_A}{V_A} - \frac{\rho_A}{\rho_c}\frac{\Delta V_c}{V_c} \tag{13}$$

where ρ_c and ρ_A are the crystal and bulk densities (2.26 and 1.74 g/cm^3), respectively.

Figure 33 compares the change in pore volume due to bromination, calculated from Eq. (13) and the data in Fig. 31, with the measured strength changes. Figure 33(a) shows that initially porosity is closed with a net decrease of about 5% (at about 15% crystal strain)

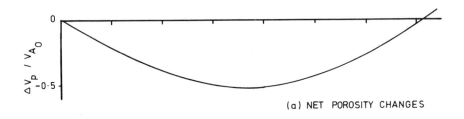

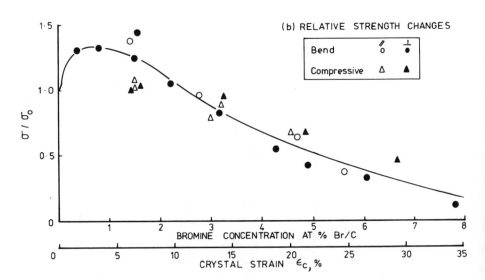

FIG. 33. Comparison of strength changes on bromination of PGA graphite and changes in porosity: (a) net porosity changes, and (b) relative strength changes.

followed by a net generation of porosity at higher crystal strains. Figure 33(b) shows the fractional change in strength with bromine content (and crystal strain) from both bend and compressive tests on the brominated PGA material. The measurements were performed rapidly after removal of the specimens from the bromine atmosphere and auxiliary determinations of the rate of bromine loss established the actual content at the time of the strength test. Figure 33 is analogous to Figs. 27 and 28, which show the irradiation-induced volume changes and strength changes of graphite. The irradiation-induced crystal volume changes are near zero under the irradiation conditions considered, hence the bulk volume changes represent the net changes in porosity. In Fig. 33(b) the strength increases initially with bromine content and then falls, following the structural changes shown in Fig. 33(a), i.e., pore closure followed by pore generation, but with the strength peak occurring well before the minimum in the net pore change. However, the interpretation of the data is far from simple since the presence of bromine has influences other than its effect on the macrostructure discussed previously.

Tsuzuku and Saito [152] have shown that low bromine concentrations increase the Young's modulus of graphite. Initial concentrations of bromine may inhibit the motion of the basal plane dislocations and increase the effective crystal shear modulus, but as crystal strain become large (55% at saturation), the effective modulus could decrease due to a lower degree of interaction between adjacent layers. Hence mechanisms for an initial increase at low concentrations and decrease at high concentrations are possible. These changes will be reflected in the Young's modulus of the polycrystal and hence also in strength. In addition, the surface energy may change, and the effective inherent crack length undoubtedly increases. The separation of these potential contributions to the strength changes is not possible from these results, but they do show that the material has some residual strength even after large dilations (several percent linear) compared with the tensile strain to failure ($\sim 0.1\%$). Unpublished work [153] has shown that on bromine removal the strength reduction is at least partially recoverable, with some hysteresis

reflecting that in dimensional changes. Further study of this phenomenon might be profitable.

XIII.   SUMMARIZING DISCUSSION

At the outset, the problem was noted of a lack of quantitative description of the *general* failure condition for polycrystalline graphite, relating ultimate failure to the applied stress state, allowing
for environmental factors and effects of time and stress cycling.
This problem is particularly important in engineering design situations, where compromise solutions are sought by defining the "strength"
of the material as near as possible under in-service conditions, and
making empirical allowances for variables. The review set out to
examine the extent of knowledge of the fracture characteristics of
polycrystalline graphite and efforts made to understand the process,
with the additional aim of identifying the current problem areas.
Of necessity, the work has been divided into sections to describe
particular aspects in detail, but these sections are all part of the
same phenomenon and must be recognized as such.

Most of the fracture studies reviewed here have been performed
on reasonably well-graphitized polycrystalline materials and only a
very small proportion of the studies refer to poor graphitic structures. Thus a general overall view of the behavior applies in the
main to the class of materials well toward the end of the range
described by Jenkins [1] as exhibiting a high degree of interlamellar
shear under an external load. Comparatively little work has been done
on materials at the other end of the scale, e.g., glassy carbons.

In the initial state, the polycrystalline material may contain
large voids together with a whole spectrum of pores down to unavoid-
able cooling cracks. When an external stress is applied, the internal
stress distribution is governed in the first instance by the size
and distribution of the larger pores. These pores concentrate the
stress and define the ultimate fracture path. The initial strain
of the well-graphitized body is dominated by the basal plane shear
of the crystals, and a certain amount of basal plane slip occurs,

limited by the elastic deformation of the surrounding material. The
crystals deform into available pore space and the Poisson's ratio
of the polycrystal is small.

When the elastic strain in a localized region, e.g., a grain,
exceeds a critical value, the growth of suitably oriented microcracks
occurs, governed by the surface energy and causing relief of internal
stress. Thus externally applied strain is absorbed by elastic and
limited plastic deformation and by microcracking, but the material
retains an excellent memory of its original state by means of a
"long chain" elastic network of relatively high compliance.

The microcracking occurs most readily as cleavage cracks along
basal planes, and propagation is restricted by cracks running either
into pores, or deviating out of critical orientation with respect to
the local stress pattern at obstacles or by following the layer structure. Thus there is a build up of crack density. This development
of nonpropagating microcracks proceeds with increasing stress, and
interlinking occurs between suitably oriented neighbors. The scale
on which this crack growth occurs will increase from near-perfect
crystallite regions within the coke grains until the grain itself is
regarded as the fundamental unit, and further interlinking will then
occur between suitably oriented grains when sufficient strain energy
is available to fracture the misoriented region at the grain boundaries. This microscopically observable crack growth then makes a
large contribution to the absorption of externally applied strain at
high stress levels, and linked cracks occur normal or at small angles
to the applied stress. In anisotropic materials, the observed cracks
are longer when the external stress is applied perpendicular to extrusion or parallel to molding due to the preferred orientation of
coke particles and the basal plane structure within them.

Ultimately, a critical crack density is achieved when the interlinking of cracks results in an effective critical length for macrofailure to occur. This critical density is first reached in regions
of high-stress concentration near the larger pores, which may contribute to an effective crack length under near uniform stress fields
by virtue of microcrack formation in diametrically opposed regions

of the large pore. Under nonuniform stress fields, these large pores
will tend to arrest cracks rather than contribute to the critical
crack size, and their role must be considered in terms of the applied
stress gradient relative to the pore size.

Qualitatively, this complex fracture process of microcrack growth
accounts for a number of the experimental fracture observations cov-
ered in this review. The effective work of fracture of the polycrystalline material is high compared to the crystal basal plane values
due to multiple crack growth, crack branching, and some contribution
from nonbasal plane fracture surfaces. The notch sensitivity is low,
since beyond a certain limit, sharpening the root of an artificial
notch produces no further effect and the microstructural features
(e.g., grain size) set the limit of stress concentration. Coarse-
grained materials tend to be less notch sensitive but are generally
weaker than their fine-grained counterparts. In the same way, small
artificial defects produce insignificant effects on strength until a
critical depth is reached which in some materials can be identified
with an effective inherent flaw size calculated via fracture mechanics,
correlating with the maximum grain and/or pore size of the material.
Some of the very fine-grained materials may be exceptions in this
respect. The exact role played by the filler grains and the large
pores in the structure will depend on the stress gradient as noted
previously, but in general, the elimination of avoidable porosity
and, for a given raw material, the use of a fine grain size will
result in a stronger product. There is, however, evidence that high-
density, extremely fine-grained materials are also more brittle.

Quantitatively, the crack mechanism suggests a Griffith type
behavior with a weak-link mechanism involved, but the complexity of
the process has presented difficulties in obtaining simple relationships between the external failure stress, the basic physical properties of graphite, and the microstructural features.

The foregoing difficulties have led to phenomenological descriptions of the material at the point of failure under specific situations, e.g., multiaxial stress, nonuniform stress, fatigue and

environmental effects, and the attempt to unify these separately
treated situations by the application of fracture mechanics in which
the material is described in terms of effective values of inherent
defect size and work of fracture.

The phenomenological studies of multiaxial stress effects appear
to be best represented for engineering purposes by an empirical equation of an energy type relationship between the principal stress
directions, although some progress has been made in describing tensile-
tensile data by Weibull statistics. Attempts to apply Weibull statistics to intercomparisons of different laboratory tensile tests
show, in a particular example, that observed strength distributions
at constant volume, volume effects, and bend/tensile strength ratios,
may be consistently satisfied by a single set of parameters provided
that allowance can be made for grain size effects. These effects are,
however, difficult to isolate and less important under stress gradients, and an adequate analytical treatment has not been formulated.
The fracture mechanics approach, which relies on the concept of an
inherent defect in a linear elastic solid, appears to: (1) give reasonably consistent values for effective toughness from different specimen types of the same material, (2) explain observed strength results
from different types of test in limited intercomparisons, and (3) give
a degree of success in producing effective inherent flaw sizes consistent with both microstructural features and experimental determinations
of effective defect dimensions. However, there is evidence here also
that different stress gradients in the test influence the effective
inherent flaw size obtained and this presents an obstacle to the
general application of fracture mechanics. Particular attention is
drawn here to the results in Figs. 7, 11, and 12, and their related
discussions in the text. The size effect studies showed grain size to
be less important in bend than in tension, and the fracture mechanics
approach showed an effective defect size which was smaller in bend
than in uniform tension. These observations are believed to be related and possibly an important lead in the further development of
a general engineering fracture criteria. A satisfactory treatment

must solve the problem of relating the applied stress gradient (usually defined for a homogeneous continuum) and the real structure,
since attempts to characterize the latter by a unique maximum defect
appear to give results which depend on the applied stress distribution.

The discussion on fatigue explored in some detail the success of
a particular empirical design model in explaining observed room temperature data obtained in air. However, very little is known about
the true mechanism involved and the influence of test atmosphere,
although it would appear that incremental crack growth (or crack
density increase) occurs on each cycle. The internally adsorbed
moisture in as-received graphite has been shown to significantly
reduce strength and work of fracture and may similarly enhance fatigue
behavior. The small amount of data available on temperature effects
on fatigue suggests that the effect is reduced at high temperatures,
which might be associated with the removal of adsorbed water. Tests
in vacuum are not available to confirm the reduction or absence of
fatigue effects under such conditions. These effects of adsorbed
moisture may also be related to those of the so-called static fatigue
effects observed when fracture occurs after long time periods in
which graphite is subjected to constant stress levels which are large
fractions of the short time failure stress.

The observed effects of varying grain size, density, crystallinity, etc., were examined against a Griffith type relationship
between the different parameters in an attempt to find a consistent
description for all the data. Allowing that modulus averages the
crack structure while strength samples the maximum crack sizes, the
following generalizations may be made.

1. Strength increases inversely proportional to the square
    root of the grain size.

2. Changes within the crystal lattice without changes in the
    basic macrostructure cause strength and modulus changes such
    that strength is proportional to the square root of the mod-
    ulus, i.e., failure occurs at constant strain energy, both
    within the grains and in the sample. Effective values of
    inherent flaw size and work of fracture are expected to
    remain constant.

3. Structural changes external to the crystal lattice, e.g.,
    by impregnation or oxidation of the same basic structure,
    lead to almost the same fractional changes in strength and
    modulus such that failure occurs at constant strain, which
    can be interpreted as indicating failure at constant strain
    energy within the grains. Effective values of inherent
    flaw size and work of fracture both change under these
    conditions.

These generalizations give a framework in which to describe and
interpret the effects, but there are detailed exceptions and interactions which must be recognized. Deviations from the constant
strain condition are observed when density variants are introduced at
early stages in the manufacturing process rather than by impregnation
or oxidation of the formed product, i.e., of the same basic structure.
The dependence of strength on density is dependent on grain size,
being a maximum for small grains, but artificial defects can over-
ride this effect when their dimensions exceed the natural defect
size determined by the grains.

Fast-neutron irradiation changes modulus and strength such that
failure occurs between the limits of constant strain and constant
strain energy, but the material type and the initial degree of
crystallite perfection influence the results. Irradiation effects
are made more complex by structural changes due to crystallite dimensional changes superimposed on the changes induced within the
crystal lattice, and the similar effects are observed on intercalation of foreign atoms. Under certain conditions, these different
contributions to strength and modulus changes can be separately
identified.

Thermal closure of cracks causes increased strength with temperature up to 2000°C, when plastic flow begins to dominate and strain
rate effects are readily observed. The latter are also observed at
room temperature when strain rates are very high, and in impact
situations, there is a basic problem of relating behavior to standard statically determined properties in absolute terms.

Empirically, many of these effects of structure variants and
environment can be allowed for, very often in terms of the fractional

changes they produce, which normalize behavior of different graphite types, e.g., oxidation effects, but the aim of the modeller must be to incorporate these effects into a satisfactory physically based model for the material behavior. Ideally, the development of such physically based models should be pursued to bridge the gap between them and the continuum behavior of the material important to the engineer, aiming to offer a description of most of the well-characterized behavior, if not all the observations, reviewed here.

A theoretical approach to fracture in a real material can only be made by idealizing the structure of the material and the applied stress pattern. Such an idealized model must describe the observed behavior and must satisfy the intercomparison of different laboratory tests before being offered to the engineer for use in general design problems. One approach which has not been applied to graphite problems but which might be worth exploring is that due to Fisher and Hollomon [154], and starts from the assumption that the material contains a statistical distribution of crack sizes. A model which has been applied to graphite with some success in its ability to link observed behavior with microstructural features has been developed comparatively recently by Meyer and Zimmer [155], based on the coincidental alignment of microcracks to achieve a critical size. The model is supported by experimental work [156], which includes excellent photomicrographs of crack progression during the failure process. This model predicts the stress–strain behavior and tensile failure point on the basis of microstructural parameters such as porosity, grain size and orientation, and the fracture toughness of the material, and can be used to indicate the changes in the microstructure necessary for improved mechanical properties.

The strength of graphite clearly is not a unique property of the material but depends on the test employed, i.e., on the sample geometry and the stress distribution, and the outstanding materials science problem is the complete interrelation of these tests, with extension to other more complex loading patterns and geometries. The most promising line of attack at present appears to be to idealize the material using a fracture mechanics description, which in principle can be used with complex geometries and stress patterns by the use of finite element techniques. This approach must incorporate a statistical treatment, perhaps using a distribution of effective flaw sizes appropriate to the applied stress distribution. It will be necessary to investigate further the dependence of effective crack size on applied stress distribution to formulate such an approach*.

Many of the general problems outlined here are not peculiar to graphite but must exist to a greater or lesser degree in other materials, although graphite has some unique features in its structure and resulting behavior. Failure in rocks and concrete, for example, have some similar characteristics to those of graphite. While there are some aspects of the fracture behavior peculiar to or high-lighted in a given material, there is much common ground and progress in understanding should be possible by combining the experiences gained in the fracture field on a wide range of materials. In pursuing this topic, the materials scientist should seek to improve his ability to satisfy his engineering colleagues on this difficult, but very common, materials problem of explaining why materials fail and defining the failure limits of components in any given situation.

ACKNOWLEDGEMENTS

The author acknowledges the encouragement and advice of his colleague Mr. B. T. Kelly in the writing of this manuscript, and the invaluable assistance of his wife in its preparation. It is published by permission of Mr. R. V. Moore, Member for Reactors, UKAEA.

---

*In a recent numerical analysis of fracture data using finite element methods, it was proposed that a failure criterion could be based on a constant K_{Ic} and a crack size which decreases as the maximum tensile strength gradient in the specimen increases [157].

REFERENCES

1. G. M. Jenkins, in *Chemistry and Physics of Carbon*, Vol. 11
   (P. L. Walker, Jr., and P. A. Thrower, eds.) Dekker, New York,
   1973, pp. 189-242.

2. G. Jouquet, *Bull. Informations Scientifiques et Techniques*,
   C.E.A., No. 192, Phenomenes de rupture dans les solides, May
   1974, p. 25.

3. H. H. W. Losty, in *Nuclear Graphite* (R. E. Nightingale, Ed.)
   Academic Press, New York, 1962, Chapters 11 and 6.

4. J. Rappeneau and G. Jouquet, *Les Carbones*, Masson, Paris, 1965,
   Chapter 11.

5. W. N. Reynolds, *Physical Properties of Graphite*, Elsevier,
   New York, 1968.

6. S. Mrozowski, in *Proc. 1st and 2nd Conferences on Carbon,
   Buffalo*, 1953/55, Waverley Press, Baltimore, 1956, p. 31.

7. P. A. Thrower and W. N. Reynolds, *J. Nucl. Mater.*, *8*, 221
   (1963).

8. M. C. Smith, in *Proc. The Conference on Continuum Aspects of
   Graphite Design, Gatlinburg, 1970* (W. L. Greenstreet and G. C.
   Battle, Jr., Eds.), USAEC Report CONF-701105, February 1972,
   p. 475.

9. C. D. Pears, *Ibid.*, p. 488.

10. C. L. Mantell, *Carbon and Graphite Handbook*, Interscience,
    New York, 1968.

11. H. H. W. Losty and J. S. Orchard, in *Proc. 5th Carbon Confer-
    ence, Penn. State, 1961*, Vol. I, Pergamon Press, New York,
    1962, p. 519.

12. E. J. Seldin, *Carbon, 4*, 177 (1966).

13. W. L. Greenstreet, J. E. Smith, and G. T. Yahr, *Carbon, 7*,
    15 (1969).

14. G. M. Jenkins, *Brit. J. Appl. Phys.*, *13*, 30 (1962).

15. G. M. Jenkins, *J. Nucl. Mater.*, *29*, 322 (1969).

16. J. H. W. Simmons, in *Proc. 3rd Carbon Conference, Buffalo,
    1957*, Pergamon Press, New York, 1959, p. 559.

17. H. W. Davidson and H. H. W. Losty, in *Proc. 4th Carbon Confer-
    ence, Buffalo, 1959*, Pergamon Press, New York, 1960, p. 585.

18. S. Amelinckx and J. Delavignette, *Phil. Mag.*, *5*, 533 (1960).

19. G. K. Williamson, *Proc. Roy. Soc. London Ser. A*, *257*, 457
    (1960).

20. C. Baker and A. Kelly, *Phil. Mag.*, *9*, 927 (1964).

21. B. T. Kelly, *Phil. Mag.*, *9*, 721 (1964).

22. W. N. Reynolds, *Phil. Mag.*, *11*, 357 (1965).

23. B. T. Kelly and A. J. E. Foreman, *Carbon, 12*, 151 (1974).

24. D. E. Soule and C. W. Nezbeda, *J. Appl. Phys., 39*, 5122 (1968).

25. O. L. Blakslee, D. G. Proctor, E. J. Seldin, G. B. Spence, and T. Weng, *J. Appl. Phys., 41*, 3373 (1970).

26. J. F. Andrew, J. Okada and D. C. Wobschall, in *Proc. 4th Carbon Conference, Buffalo, 1959*, Pergamon Press, New York, 1960, p. 559.

27. K. E. Gilchrist and D. Wells, *Carbon, 7*, 627 (1969).

28. G. M. Jenkins, *J. Nucl. Mater., 5*, 280 (1962).

29. O. D. Slagle, *J. Amer. Ceram. Soc., 50* (10), 495 (1967).

30. R. Taylor, R. G. Brown, K. E. Gilchrist, E. Hall, A. T. Hodds, B. T. Kelly, and F. Morris, *Carbon, 5*, 519 (1967).

31. R. H. Knibbs, *J. Nucl. Mater., 24*, 174 (1967).

32. T. Oku and M. Eto, *Carbon, 11*, 639 (1973).

33. R. A. Meyer and J. D. Buch, *11th Carbon Conference Abstracts*, CONF-730601, 229 (1973).

34. W. V. Green, L. S. Levinson, R. D. Reiswig, and E. G. Zukas, *Carbon, 5*, 583 (1967).

35. E. G. Zukas and W. V. Green, *Carbon, 10*, 519 (1972).

36. W. V. Green, E. G. Zukas, and J. Weertman, *Trans. ASM, 62*, 512 (1969).

37. E. Hall, *J. Nucl. Mater., 15*, 137 (1965).

38. P. E. Hart, *Carbon, 10*, 233 (1972).

39. A. A. Griffith, *Phil. Trans. Roy. Soc. A, 221*, 163 (1920).

40. M. Eto and T. Oku, *J. Nucl. Mater., 54* (2), 245 (1974).

41. R. B. Matthews, *J. Amer. Ceram. Soc., 57* (5), 225 (1974).

42. J. W. Dally and L. N. Hjelm, *J. Amer. Ceram. Soc., 48* (7), 338 (1965).

43. J. Jortner, in *Proc. The Conference on Continuum Aspects of Graphite Design, Gatlinburg, 1970* (W. L. Greenstreet and G. C. Battle, Jr., Eds.), USAEC Report CONF-701105, February 1972, p. 641.

44. B. J. S. Wilkins, *J. Amer. Ceram. Soc., 54*, 593 (1971).

45. P. H. Hodkinson and J. S. Nadeau, *J. Mater. Sci., 10*, 846 (1975).

46. R. G. Bruce, *J. Metal. Club, R. C. S. Glasgow*, No. 10, 41 (1958-1959).

47. R. J. Good, L. A. Girifalco and G. Kraus, *J. Phys. Chem., 62*, 1418 (1958).

48. von Mises, *Z. Angew. Math. Mech., 8*, 161 (1928).

49. G. W. Groves and A. Kelly, *Phil. Mag.*, *8*, 877 (1963).

50. I. B. Mason, in *Proc. 5th Carbon Conference, Penn. State, 1961*, Vol. II, Pergamon Press, Oxford, 1963, p. 597.

51. Yu S. Virgil'ev, E. I. Kurolenkin, V. G. Makarchenko, and T. K. Pekal'n, *Strength Mater.* (Engl. Transl.), *5* (11), 1336 (1973).

52. N. J. Petch, *J. Iron Steel Inst.*, *174*, Part I, 25 (1953).

53. J. D. Buch, *11th Carbon Conference Abstracts*, CONF-730601, 231 (1973).

54. C. A. Anderson and E. I. Salkovitz, *ibid.*, 233 (1973).

55. A. A. Griffith, in *Proc. 1st Int. Congress Appl. Mech.*, *Delft, 1924*, p. 55.

56. S. A. F. Murrell, *Brit. J. Appl. Phys.*, *15*, 1195 (1964).

57. F. E. McClintock and J. B. Walsh, in *Proc. 4th U.S. Congress Appl. Mech., Berkeley, 1962*, Amer. Soc. Mech. Engs. New York, 1963, p. 1015.

58. H. W. Babel and G. Sines, *J. Basic Engr.*, 285 (June 1968).

59. W. J. Weibull, *Appl. Mech.*, *18*, 293 (1951).

60. L. Obert, in *Fracture, An Advanced Treatise*, Vol. VII (Fracture of Non-metals and Composites) (H. Liebowitz, Ed.), Academic Press, New York, 1972, Chapter 3, p. 93.

61. W. F. Brace and E. G. Bombolakis, *J. Geophys. Res.*, *68*, 3709 (1963).

62. L. M. Gillin, *J. Nucl. Mater.*, *23*, 280 (1967).

63. W. G. Bradshaw, in *Proc. The Conference on Continuum Aspects of Graphite Design, Gatlinburg, 1970* (W. L. Greenstreet and G. C. Battle, Jr., Eds.) USAEC Report CONF-701105, February 1972, p. 40.

64. R. E. Ely, *J. Amer. Ceram. Soc.*, *48* (10), 505 (1965).

65. R. E. Ely, *Ceram. Bull.*, *47* (5), 489 (1968).

66. L. J. Broutman, S. M. Krishnakumar, and P. K. Mallick, *J. Amer. Ceram. Soc.*, *53* (12), 649 (1970).

67. J. Jortner, McDonnell Douglas Astronautics Company, Santa Monica, California, Technical Report AFML-TR-71-253, December 1971.

68. J. Jortner, in *Proc. The Conference on Continuum Aspects of Graphite Design, Gatlinburg, 1970* (W. L. Greenstreet and G. C. Battle, Jr., Eds.), USAEC Report CONF-701105, February 1972, p. 514.

69. Tu-Lung Weng, *ibid.*, p. 222.

70. S. G. Babcock, S. J. Green, P. A. Hochstein and J. A. Gum, *ibid.*, p. 50.

71. R. J. Price and H. R. W. Cobb, *ibid.*, p. 547.

72. S. B. Batdorf and J. C. Crose, *J. Appl. Mech. (Trans. ASME)*, 459 (June 1974).

73. J. C. Jaeger, *Elasticity, Fracture and Flow*, 3rd Ed., Metheun, 1969.

74. E. A. Kmetko, J. R. Morgan and J. R. Andrew, *Carbon, 6*, 571 (1968).

75. M. S. Paterson and J. M. Edmond, *Carbon, 10*, 29 (1972).

76. J. E. Brocklehurst and M. I. Darby, *Mater. Sci. Engr., 16*, 91 (1974); *18*, 304 (1975).

77. J. Amesz, J. Donea, and F. Lanza, *11th Biennial Conference on Carbon, Gatlinburg, 1973, Extended Abstracts*, CONF-730601, p. 221.

78. R. L. Wooley, *Phil. Mag., 11*, 799 (1965).

79. V. Lungagnani and R. Krefeld, in *Proc. The Conference on Continuum Aspects of Graphite Design, Gatlinburg, 1970* (W. L. Greenstreet and G. C. Battle, Jr., Eds.) USAEC Report CONF-701105, February 1972, p. 663.

80. F. Lanza and H. Burg, *11th Biennial Conference on Carbon, Gatlinburg, 1973, Extended Abstracts*, CONF-730601, p. 223.

81. W. B. Powell and P. E. Massier, in *Proc. 3rd Carbon Conference, Buffalo, 1957*, Pergamon Press, New York, 1959, p. 543.

82. R. W. Andrae, in *Proc. The Conference on Continuum Aspects of Graphite Design, Gatlinburg, 1970* (W. L. Greenstreet and G. C. Battle, Jr., Eds.), USAEC Report CONF-701105, February 1972, p. 490.

83. H. G. Tattersall and G. Tappin, *J. Mater. Sci., 1*, 296 (1966).

84. R. W. Davidge and G. Tappin, *J. Mater. Sci., 3*, 165 (1968).

85. F. H. Vitovec, *J. Test. Evaluat., 1* (3), 250 (1973).

86. A. L. Udovskii, N. O. Gussman, and V. N. Barabanov, *Strength Mater., 4* (5), 595 (1972).

87. J. M. Corum, *J. Nucl. Mater., 22*, 41 (1967).

88. G. T. Yahr and R. S. Valachovic, in *Proc. The Conference on Continuum Aspects of Graphite Design, Gatlinburg, 1970* W. L. Greenstreet and G. C. Battle, Jr., Eds.), USAEC Report CONF-701105, February 1972, p. 533.

89. G. T. Yahr, R. S. Valachovic, and W. L. Greenstreet, paper in Graphite Structures for Nuclear Reactors Conference, organized by the Nuclear Energy Group of the Institution of Mechanical Engineers, March 1972.

90. J. E. Srawley and W. F. Brown, *Fracture Toughness Testing and its Applications*, ASTM Spec. Tech. Publication No. 381, 1965, p. 30.

91. E. T. Wessel, *Eng. Fracture Mech., 1*, 77 (1968).

92. P. Marshall and E. K. Priddle, *Carbon, 11*, 541 (1973)

93. F. H. Vitovec and Z. H. Stachurski, *Carbon, 10*, 417 (1972).

94. R. Stevens, *Carbon, 9*, 573 (1971).

95. J. R. Dixon and J. S. Strannigan, *J. Strain Anal., 7*, 125 (1972).

96. P. Marshall and E. K. Priddle, *Carbon, 11*, 627 (1973).

97. L. Green, *J. Appl. Mech., 18*, 345 (1951).

98. V. N. Barabanov, Yu. P. Anufriev, G. G. Zaitsev, and M. Ya. Pimkin, *Zavod. Lab., 32*, 4, 459 (1966).

99. S. Sato, Y. Imamura, K. Kawamata, J. Kon, and M. Ohtani, *J. Soc. Mater. Sci., Japan, 20* (210), 75 (1971).

100. H. L. Leichter and E. Robinson, *J. Amer. Ceram. Soc., 53* (4), 197 (1970).

101. B. J. S. Wilkins, *J. Materials, 7* (2), 251 (1972).

102. B. J. S. Wilkins and A. R. Reich, AECL-4216 (1972).

103. B. J. S. Wilkins and B. F. Jones, *J. Mater. Sci., 8*, 1362 (1973).

104. H. Neuber, in *Advanced Mechanics of Materials* (Seely and Smith, Eds.), 2nd Ed., Wiley, New York, 1952, p. 391.

105. R. B. Heywood, *Designing by Photoelasticity*, Chapman and Hall, London, 1952.

106. R. E. Peterson, *Stress Concentration Design Factors*, Wiley, New York, 1953.

107. M. O. Tucker, C E G B Berkeley Nuclear Laboratories, UK, to be published.

108. D. K. Bazaj and E. E. Cox, *Carbon, 7*, 689 (1969).

109. J. M. Hutcheon and M. S. T. Price, in *Proc. 4th Carbon Conference, Buffalo, 1959*, Pergamon Press, Oxford, 1960, p. 645.

110. N. Hawkins, in *Proc. 2nd Conference on Industrial Carbon and Graphite*, Society of Chemical Industry, London, 1966, p. 355.

111. J. E. Brocklehurst, R. G. Brown, K. E. Gilchrist, and V. Y. Labaton, *J. Nucl. Mater., 35*, 183 (1970).

112. J. A. Board and R. L. Squires, in *Proc. 2nd Conference on Industrial Carbon and Graphite*, Society of Chemical Industry, London, 1966, p. 289.

113. C. Rounthwaite, G. A. Lyons, and R. A. Snowdon, *ibid.*, p. 299.

114. R. H. Knibbs and J. B. Morris, in *Proc. 3rd Conference on Industrial Carbon and Graphite*, Society of Chemical Industry, London, 1971, p. 297.

115. P. E. Armstrong, in *Proc. The Conference on Continuum Aspects of Graphite Design, Gatlinburg, 1970* (W. L. Greenstreet and G. C. Battle, Jr., Eds.), USAEC Report CONF-701105, February 1972, p. 3.

116. J. Amesz and G. Volta, *11th Carbon Conference Abstracts*, CONF-730601, 225 (1973).

117. C. Malmstrom, R. Keen and L. Green, Jr., *J. Appl. Phys., 22,* 593 (1951).

118. H. E. Martens, L. D. Jaffe, and J. E. Jepson, in *Proc. 3rd Carbon Conference, Buffalo, 1957*, Pergamon Press, Oxford, 1959, p. 529.

119. R. E. Nightingale, *Nuclear Graphite*, Academic Press, New York, 1962, p. 153.

120. H. E. Martens, D. D. Button, D. B. Fischback, and L. D. Jaffe, in *Proc. 4th Carbon Conference, Buffalo, 1959*, Pergamon Press, Oxford, 1960, p. 511.

121. A. L. Sutton and V. C. Howard, *J. Nucl. Mater., 7* (1), 58 (1962).

122. J. H. W. Simmons and W. N. Reynolds, *Inst. of Metals Monograph on Uranium and Graphite*, Paper No. 11 (1962).

123. I. B. Mason and R. H. Knibbs, *Carbon, 5,* 493 (1967).

124. J. F. Andrew and S. Sato, *Carbon., 1,* 225 (1964).

125. H. W. Davidson, H. H. W. Losty, and A. M. Ross, in *Proc. 1st Conference on Industrial Carbon and Graphite*, Society of Chemical Industry, London, 1958, p. 551.

126. R. J. Diefendorf, in *Proc. 4th Carbon Conference, Buffalo, 1959*, Pergamon Press, Oxford, 1960, p. 489.

127. G. W. Rowe, *Nuc. Eng., 7,* 102 (1962).

128. D. H. Logsdail, AERE Report 5721 (1968).

129. R. J. Diefendorf, in *Proc. 4th Carbon Conference, Buffalo, 1959*, Pergamon Press, Oxford, 1960, p. 483.

130. W. N. Reynolds, *Physical Properties of Graphite*, Elsevier, New York, 1968, p. 56.

131. C. D. Pears, in *Proc. The Conference on Continuum Aspects of Graphite Design, Gatlinburg, 1970* (W. L. Greenstreet and G. C. Battle, Jr., Eds.) USAEC Report CONF-701105, February 1972, p. 115.

132. M. C. Smith, *Carbon, 1,* 147 (1964).

133. M. C. Smith, *Carbon, 2,* 269 (1964).

134. B. T. Kelly, D. Jones, and A. James, *J. Nucl. Mater., 7,* 279 (1964).

135. W. V. Green, J. Weertman, and E. G. Zukas, *Mater. Sci. Engr.,*
     6 (3), 199 (1970).

136. *Industrial Graphite Engineering Handbook*, Union Carbide Cor-
     poration, 1959.

137. P. M. Jones, in *2nd International Conference on Structural*
     Mechanics in Reactor Technology, Berlin, 1973, Vol. 1,
     Part C-D, Paper D 4/6. CID Publications, Luxembourg.

138. S. Sato and S. Miyazono, *Carbon, 2*, 103 (1964).

139. D. P. H. Hasselman, *J. Amer. Ceram. Soc., 52*, 600 (1969).

140. W. K. Woods, L. P. Bupp, and J. F. Fletcher, in *Proc. 1st*
     International Conference on the Peaceful Uses of Atomic
     Energy, Geneva, 1955, Vol. 7, p. 455. United Nations, New York (1956).

141. H. W. Davidson and H. H. W. Losty, in *Proc. 2nd United N.*
     International Conference on the Peaceful Uses of Atomic
     Energy, Geneva, 1958, Vol. 7, p. 307, United Nations, Geneva (1950).

142. P. A. Platonov, Yu. S. Virgil'ev, V. I. Karpukhin, A. L. Zaitsev,
     and I. F. Novobratskaya, *Sov. Atom. Energy, 35* (3), 805 (1973).

143. G. B. Engle, R. J. Price, W. R. Johnson, and L. A. Beavan,
     presented at 4th London International Carbon and Graphite
     Conference, Society of Chemical Industry, 1974, paper 20.

144. P. T. Nettley, J. E. Brocklehurst, W. H. Martin, and J. H. W.
     Simmons, IAEA Symp. Advanced High Temperature Gas Cooled
     Reactors, Paper SM-111/34, 1969, p. 604 (IAEA Vienna).

145. B. T. Kelly, W. H. Martin and P. T. Nettley, *Phil. Trans. Roy.*
     Soc. A, 260, 37 (1966).

146. B. T. Kelly, and J. E. Brocklehurst, *Carbon, 9*, 783 (1971).

147. B. T. Kelly and J. E. Brocklehurst, in *Proc. 3rd Conference on*
     Industrial Carbon and Graphite, Society of Chemical Industry,
     London, 1971, p. 363.

148. A. R. Ubbelohde and F. A. Lewis, *Graphite and its Crystal*
     Compounds, Oxford Univ. Press, Oxford, 1960.

149. J. E. Brocklehurst and J. C. Weeks, *J. Nucl. Mater., 9, 197*
     (1963).

150. J. E. Brocklehurst and R. A. Bishop, *Carbon, 2*, 27 (1964).

151. J. E. Brocklehurst, *Nature, 194* (4825), 247 (1962).

152. T. Tsuzuku and M. Saito, *J. Appl. Phys. Japan, 6*, 54 (1967).

153. A. Turnbull and P. Horner, Unpublished work at CEGB, Berkeley
     Nuclear Laboratories, UK.

154. J. C. Fisher and J. H. Holloman, Amer. Inst. of Mining and
     Metallurgical Engineers, Tech. Pub. No. 2218, August 1947.

155. R. A. Meyer and J. E. Zimmer, Aerospace Report No. ATR-74
      (7425)-3, 1974.

156. R. A. Meyer, J. E. Zimmer and M. C. Almon, Aerospace Report
      No. ATR-74 (7408)-2, 1974.

157. M. I. Darby, *Int. J. Fracture, 12,* 745 (1976).

Numbers in parentheses are reference numbers and indicate that an author's work is referred to although his name is not cited in the text. Underlined numbers give the pages on which the complete references are listed.

Ackerman, M. W., 23(80), 132
Agrawal, B. K., 23, 132
Aka, E. Z., 5, 129
Aksenkov, V. K., 110(283), 142
Allsopp, H. L., 2(2), 128
Almon, M. C., 270(156), 279
Amelinekx, S., 151, 272
Amesz, J., 184(77), 230, 275, 277
Anastassakis, E., 33(139), 37(139), 44(139), 52(163), 53(163), 135, 136
Anderson, C. A., 165, 274
Andrae, R. W., 192, 275
Andreatch, P., Jr., 67(214), 69(214), 138
Andrew, J. F., 153, 157(26), 232, 273, 277
Angress, J. F., 24(94), 52(163, 164), 53(163), 112(297), 132, 136, 143, 178(74), 275
Ångström, A. J., 9(30), 130
Angus, J. C., 30(115), 133
Anufriev, Yu. P., 205(98), 276
Armstrong, P. E., 299, 277
Augustynick, W. M., 106(276), 141
Austin, I. G., 31(119), 134

Babcock, S. G., 274
Babel, H. W., 168, 174, 274
Bagguley, D. M. S., 33(138), 135
Baker, C., 151, 153(20), 272
Barabanov, V. N., 195, 203(86), 205(98), 235)86), 236(86), 275, 276

Barker, M., 36(132), 134
Batdorf, S. B., 176, 275
Bazaj, D. K., 224, 239, 276
Beavan, L. A., 245(143), 278
Berman, R., 23(80,84), 24(85), 132
Bezrukov, G. N., 21(77), 22(77), 29(110), 30(115), 107(279), 131, 133, 141
Biddy, D. M., 4(3,18), 5(3), 49(3), 128, 129
Bilz, H., 51(161), 136
Birman, J. L., 54(179), 136
Bishop, R. A., 259(150), 260(150), 278
Blackwell, D. E., 9(25), 10(25), 14(25), 15(25), 50(25), 112(25), 129
Blakslee, O. L., 153, 272
Board, J. A., 228, 276
Bochko, V. A., 19(58), 131
Bohm, D., 86(253), 140
Bombolakis, E. G., 168, 274
Bond, W. L., 5(5), 10(5), 12(5), 14(5), 15(5), 19(5), 128
Borer, W. J., 52(174), 53(174), 136
Borik, M., 6(21), 23(21), 34(21), 37(21), 47(21), 50(21), 54(21), 64(21), 129
Born, M., 54(178), 136
Bovenkerk, H. P., 30(113), 36(113), 47(113), 108(113), 133
Bower, H. J., 29(109), 133
Boyle, R., 111(287), 142
Brace, W. F., 168, 274

Bradshaw, W. G., 170(63), <u>274</u>

Brafman, O., 52(173), 53(173), <u>136</u>

Braga, C. L., 99(267), 120(267), <u>141</u>

Bratashevskii, Yu. A., 110(283), <u>142</u>

Brik, A. B., 27(98), 110(283), <u>132</u>, <u>142</u>

Brocklehurst, J. E., 183, 184(76), 193(76), 228, 248(144), 252(146, 147), 256, 259(149,150,151), 260(149, 150), <u>275</u>, <u>276</u>, <u>278</u>

Brophy, J. E., 31(122), 32(122), <u>134</u>

Broutman, L. J., <u>274</u>

Brown, R. G., 155(30), 163(30), 169(30), 170(30), 174(30), 176(30), 228(111), 244(30), 256(111), <u>273</u>, <u>276</u>

Brown, W. F., 196, <u>275</u>

Bruce, R. G., 161(46), 234(46), <u>273</u>

Brust, D., 57(191), <u>137</u>

Buch, J. D., 156, 164(53), <u>273</u>, <u>274</u>

Bukhan'ko, F. N., 110(283), <u>142</u>

Bundy, F. P., 30(113), 36(113), 47(113), 108(113), <u>133</u>

Bunting, E. N., 5, <u>129</u>

Bupp, L. P., 243(140), <u>278</u>

Burg, H., 185, <u>275</u>

Burger, A. J., <u>2</u>(2), <u>128</u>

Burstein, E., 6(20), 50(20), 51(20), 52(162,163), <u>129</u>, <u>136</u>

Burton, B., 117(294), <u>142</u>

Button, D. D., 232(120), <u>277</u>

Butuzov, V. P., 24(88), 25(88), 29(110), 30(115), 107(279), <u>132</u>, <u>133</u>, <u>141</u>

Caner, M., 28(103), <u>133</u>

Cannon, P., 24(87), <u>132</u>

Carlson, C. M., 28(108), <u>133</u>

Carter, T., 5(11), 49(11), 50(11), <u>129</u>

Caticha-Ellis, S., 20(61), <u>131</u>

Caveney, R. J., 22(78), <u>132</u>

Charette, J. J., 24(86), 35(127), 37(135), <u>132</u>, <u>134</u>

Chaumet, 102(269), <u>141</u>

Chen, R., 108(282), <u>110</u>(282, 285), <u>142</u>

Chesley, F. G., 5, <u>129</u>

Chrenko, R. M., 24(89,95), 25, 27(89), 30(114), 31(117), 45(117), 46(114,117), <u>132</u>, <u>133</u>, <u>134</u>

Clark, C. D., 9, 16, 40(137), 43(140), 48(48), 49(48,152) 55-57(140,180,181), 68(48) 71(219), 74(48), 75(26), 76(229), 77(234), 78(233), 82(229), 83)229), 88(219), 90(219), 91(219,229), 94(26), 100(26), 115(233), 118(152), 119(48), 121(48,219), <u>129</u>, <u>130</u>, <u>135</u>, <u>136</u>, <u>138</u>, <u>139</u>

Clark, D., 76(227), <u>139</u>

Clegg, P. E., 33(124), 34(124), 43(124), <u>134</u>

Cleland, J. W., 68(108), <u>138</u>

Clements, W. R. L., 52(175), 53(175), <u>136</u>

Cobb, H. R. W., 174, 183, <u>274</u>

Cochran, W., 20(61), <u>131</u>

Cohen, M. H., 54(179), <u>136</u>

Collins, A. T., 31(121), 32(121), 34, 35(128), 36(128,130), 37(121,130,136), 38(130), 39(130,136), 40(136), 41, 45(121), 45(121), 47(128, 136,145,146,149,151), 75(226), 99(266), 100(266), 108(281), 110(281), <u>134</u>, <u>135</u>, <u>139</u>, <u>141</u>, <u>142</u>

Cook, R. J., 27(99), <u>132</u>

Cooke, C., 53(163), <u>136</u>

Corbett, J. W., 68(208), <u>138</u>

Corum, J. M., 195, 203, <u>275</u>

Cosk, E. E., 224, 239, <u>276</u>

Coulson, C. A., 82(249), <u>140</u>

Cowley, R. A., 50(154,156), 51(154), 54(156), 55(154), <u>135</u>, <u>136</u>

Crawfird, J. H., 68(108), <u>138</u>

Crawford, R. K., 52(173), 53(173), <u>136</u>

Crookes, W., 111(290), <u>142</u>

Crose, J. C., 176, <u>275</u>

Crossfield, M., 77(232), 77(232),
  99(266), 100(262,266),
  102(262), 107(232), 108(232),
  113(232), 115(232), 117(232),
  139, 141
Crowther, P. A., 36, 47,
  71(202), 72, 134, 137
Custers, J. F. H., 17(56),
  24(91), 31(122), 32(122),
  131, 132, 134

Dally, J. W., 160, 273
Daniels, W. B., 52(173), 53(173),
  136
Darby, M. I., 182, 184(76),
  193(76), 275
Darrow, K. A., 24(95), 132
d'Aubigne, Y. M., 71(212), 138
Davey, A. R., 67(213), 138
Davidge, R. W., 194, 275
Davidson, H. W., 151, 232,
  243(141), 272, 277, 278
Davies, G., 9(24), 11(24),
  14(45), 15(47), 17(47),
  19(47), 55(184), 62(101),
  66(201), 68)204), 69)204),
  72(47), 76(231), 77(231,
  234), 78(234), 79(234),
  81(243), 83(250), 84(201),
  85(250), 88(47,255), 90(201,
  203), 94-97(231,263), 99(266),
  100(262,266), 102(231,262),
  103(204), 105(204,231),
  108(280), 115(231,234),
  116(255), 117(231,255),
  119(255), 120(204), 123(47),
  129, 130, 137, 139, 140, 141,
  142
Dean, P. J., 5(6), 11, 14(6),
  17(49), 29(49,111), 31(116),
  36-39(128,130,131), 43-45(116,
  140,142,183), 47(111,116,128,
  131,147), 49(6,49), 54(181),
  55(116,140,183), 56(116,140),
  57(49,140), 64, 71(202), 72,
  77(235), 107(111,278),
  108(111), 113(235), 114(131),
  115(235), 118(235), 119(235),
  123(116,140), 128, 130, 133,
  134, 135, 136, 137, 139, 141

Delavignette, J., 151, 272
Denham, P., 17(49), 29, 47(148,
  151), 49(49,148), 57(49),
  130, 135
Dennison, D. M., 85(252), 140
de Sa, E., 91(256b), 92(256b),
  140
Dexter, D. L., 99(268), 101(268),
  141
Diefendorf, R. J., 233, 235,
  277
Ditchburn, R. W., 16(48), 48(48),
  49(48), 68(48), 74(48),
  99(152), 118(152), 119(48),
  121(48), 130, 135
Dixon, J. R., 202(95), 276
Dolling, G. 50(154,156), 51(154),
  54(156), 55(154), 135, 136
Donea, J., 184(77), 275
Dudenkov, Yu. A., 9(27), 24(90),
  30(90), 49(27), 108(90), 129,
  132
du Preez, L., 8(23), 24-27(23,
  97), 30(23), 68(205,206,207),
  88(207), 102(205,206), 113-
  115(207), 120(207, 129, 132,
  137
Duran, J., 71(212), 138
Dyer, H. B., 8(23), 16(48),
  24-27(23), 30(23), 48(48),
  49(48), 68(48,205,206),
  74(48), 75(225), 99(152),
  102(205,206), 118(152),
  119(48), 121(48), 129, 130,
  135, 137, 139

Edmond, J. M., 179, 275
Elliott, R. J., 21(72), 71(219),
  88(219), 90(219), 91,
  121(219), 131, 138
Ely, R. E., 175, 274
Engle, G. B., 245(143), 278
Englman, R., 28(103), 81(240),
  133, 139
Erasmus, C. S., 4(3,18), 5(3),
  49(3), 128, 129
Eto, M., 155, 158, 160(40), 273
Evans, E. L., 36(132), 134

Evans, T., 13(43), 14(46), 19,
  20(43,62,63,70), 21(43,46),
  22(46), 93(259), 130, 131,
  140
Every, A. G., 28(107), 29(110),
  133

Farrer, R. G., 28, 29, 76(228,
  271), 82(247), 103, 105,
  120(247), 133, 139, 140, 141
Faulkner, E. A., 24(93), 132
Faulkner, R. A., 44(142), 135
Ferdinando, P., 75(225), 139
Fasq, H., 4(3), 5(3), 49(3), 128
Fetterman, H. R., 83, 140
Fischback, D. B., 232(120), 277
Fisher, J. C., 270, 278
Fitchen, D. B., 60(197), 83, 137,
  140
Fletcher, J. F., 243(140), 278
Foreman, A. J. E., 152(23), 272
Foster, E. L., 23(80), 132
Fox, J. J., 9(36,37), 16(36),
  17(36), 51(37), 52(36,37),
  57(36), 130
Frank, A., 111(286), 142
Frank, F. C., 20, 21(74), 131
Freedman, M. S., 5(16), 129
Friedel, J., 82(249), 140
Frosch, C. J., 106(275), 141
Futergendler, S. I., 21(76),
  22(76), 131

Gaisin, V. A., 82(246), 90(258),
  92(258), 140
Ganeson, S., 52(162), 136
Gardner, N. C., 30(115), 133
Garzon, O. L., 5(9), 129
Gebhardt, W., 58(193), 137
Geick, R., 51(161), 136
Gelles, I. L., 27(96), 28(96),
  132
Gerasimenko, N. N., 30(115),
  107(279), 133, 141
Gerlich, D., 23(81), 132
Giardini, A. A., 5(10), 129
Giesecke, P., 60(199), 137
Gilchrist, K. E., 153, 155(30),
  163(30), 169(30), 170(30),
  174(30), 176(30), 228(111),
  244(30), 256(111), 273, 276

Gillin, L. M., 170, 235(62),
  238, 274
Girifalco, L. A., 161(47),
  234(47), 273
Glover, G. H., 46(144), 135
Glynn, P., 67(214), 69(214),
  138
Good, R. J., 161(47), 234(47),
  273
Goodwin, A. R., 6(21), 23(21),
  34(21), 37(21), 47(21),
  50(21), 54(21), 64(21),
  112(297), 129, 143
Gora, T., 36(132), 134
Green, L., 205, 231(117), 235,
  276, 277
Green, S. J., 274
Green, W. V., 156(34), 157(35,
  36), 239(34,35,36,135), 273,
  278
Greenstreet, W. L., 150, 196(89),
  204(89), 272, 275
Grenville-Wells, H. J., 20(68),
  131
Griffith, A. A., 160, 161, 166,
  180, 192, 273, 274
Groves, G. W., 161, 273
Gudden, B., 9, 16(33), 17(33),
  130
Gugenheim, H. J., 37(134), 134
Gum, J. A., 274
Gussman, N. O., 195, 203(86),
  235(86), 236(86), 275

Hall, E., 155(30), 158, 163(30),
  169(30), 170(30), 174(30),
  176(30), 244(30), 273
Halperin, A., 16(51), 17(51),
  60(200), 67, 106(277),
  111(285), 130, 137, 141, 142
Ham, F. S., 69(209,210), 71(210),
  80(209,210), 81(210), 138, 139
Hamer, M. F., 83(250), 85(250),
  90(256a), 140
Hardy, J. R., 35, 37(135), 51,
  134, 136
Harris, J. W., 4(17), 129
Harris, P. V., 43(140), 55-57(140),
  123(140), 135
Hart, P. E., 158, 273
Hasselman, D. P. H., 242, 278
Hawkins, N., 228, 276

Henderson, B., 60(198), 61(198), 137
Henry, C. H., 106(275), 141
Henvis, B. W., 40(137), 135
Heywood, R. B., 222(105), 276
Hjelm, L. N., 160, 273
Hochstein, P. A., 274
Hodds, A. T., 155(30), 163(30), 169(30), 170(30), 174(30), 176(30), 244(30), 273
Hodkinson, P. H., 160, 273
Hoerni, J. A., 20(69), 131
Holloman, J. H., 270, 278
Hopfield, J. J., 41(141b), 106(276), 135, 141
Horner, P., 263(153), 278
Howard, V. C., 232(121), 277
Huang, K., 54(178), 136
Huggins, C. M., 24(87), 132
Hughes, A. E., 90(224), 138
Hutcheon, J. M., 226, 276

Illegems, M., 44(142), 135
Ill'in, V. E., 24(88), 25(88), 118(295), 132, 142
Imamura, Y., 205(99), 276
Iwasa, S., 52(163), 53(163), 136

Jaeger, J. C., 177(73), 275
Jaffe, L. D., 231(118), 232(120), 277
James, A., 238(134), 277
James, P. F., 20(62), 131
Jenkins, G. M., 146, 150, 151, 152, 154, 159, 164, 233, 264, 271, 272, 273
Jepson, J. E., 231(118), 277
Johnson, F. A., 51, 136
Johnson, W. R., 245(143), 278
Jones, B. F., 218, 276
Jones, D., 238(134), 277
Jones, I. H., 45(183), 55(183), 137
Jones, P. M., 240, 244, 278
Jortner, J., 160, 171(67,68), 172(67,68), 175, 176, 179, 273, 274
Jouquet, G., 147, 155, 271, 272
Julius, W. H., 9, 130

Kable, E. J., 4(3), 5(3), 49(3), 128
Kaiser, W., 5(5), 10(5), 12(5), 14(5), 15(5), 19(5), 52(177), 128, 136
Kalnajs, J., 19(57,59), 131
Kamiya, Y., 24(92), 132
Kane, E. O., 35, 134
Kaplyanskii, A. A., 71(218), 90(257), 92(257), 138, 140
Karpukhin, V. I., 244(142), 278
Kawamata, K., 205(99), 276
Keen, R., 231(117), 277
Keil, T. H., 60(194), 61(194), 137
Kelly, A., 151, 153(20), 161, 272, 273
Kelly, B. T., 151, 152(23), 155(30), 163(30), 169(30), 170(30), 176(30), 238(134), 243(30), 251(145), 252(146, 147), 272, 273, 277, 278
Kemmey, P., 40(137), 135
Kim, Y. M., 110(283), 142
Kimura, S., 44(142), 135
Kiraki, A., 68(208), 138
Klemens, P. G., 22(79), 23(79, 80), 132
Klick, C. L., 17(55), 131
Klyuev, Yu. A., 9, 24(90), 30(90), 49(27), 108(90), 129, 132
Kmentko, E. A., 178, 275
Knibbs, R. H., 155, 229, 230, 232(123), 233, 273, 276, 277
Kohn, W., 32(123), 36(123), 37(123), 46(123), 134
Kolyshkin, V. I., 71(218), 138
Kon, J., 205(99), 276
Konorova, E. A., 17(52), 57(52), 130
Krasnitsa, V. N., 99(265), 141
Kraus, G., 161(47), 234(47), 273
Krefeld, R., 184, 255, 275
Kress, W., 6(21), 23(21), 34(21), 37(21), 47(21), 50(21), 54(21), 64(21), 129
Krishnakumar, S. M., 274
Krishnamurti, D., 52(167), 136
Krishnan, R. S., 52(171,165), 54(165), 67(213), 136, 138

Kristianpoller, N., 108(282), 110(282), _142_
Kühnert, H., 58(193), _137_
Kurolenkin, E. I., 163(151), _274_

Labaton, V. Y., 228(111), 256(111), _276_
Laner, J. N., 110(283), _142_
Lang, A. R., 20(71), 21(73,75), 24(92), 93)259), _131_, _132_, _140_
Lannoo, M., 82(248,249), _140_
Lanza, F., 184(77), 185, _275_
Larkins, F. P., 28(108), 80(237, 238), 82(249), _133_, _139_, _140_
Lasher, G. J., 27(96), 28(96), _132_
Laubereau, A., 52(177), _136_
Lax, M., 6(20), 50(20), 51(20), _129_
Leichter, H. L., 205, _276_
Leman, G., 82(249), _140_
Lenskaya, S. V., 9(28a), 13(42), 20(28a), 21(28a), 24(88), 25(88), _130_, _132_
Lenz, H., 17(54), _131_
Levi, M., 9, _130_
Levinson, L. S., 156(34), 239(34), _273_
Lewis, F. A., 258, _278_
Lezheiko, L. V., 30(115), 107(279), _133_, _141_
Lidiard, A. B., 27(100), 80(238), 82(249), _132_, _139_, _140_
Lightowlers, E. C., 5(6,7,8), 11, 14(6), 17(49), 29(49), 31(116), 35(128), 36(128,130), 37(130,136), 38(130), 39(130, 136), 40(136), 41, 43-47(86, 116,128,136,143,145,146,148, 149,151), 49(6,49,148), 55(116), 56(116), 57(49), 68(204), 69(204), 77(235), 99(266), 100(266), 103(204), 105(204), 108(281), 110(281), 113(235), 118(235), 119(235), 120(204), 123(116), _128-130_, _134_, _135_, _137_, _139_, _141_, _142_
Linde, von der, D., 52(177), _136_
Lisoyvan, V. I., 9(28a), 11, 13(42), 18(41), 19, 20(28a), 21(28a), _130_

Litvin, Y. A., 24(88), 25(88), 30(115), 107(279), _132_, _133_, _141_
Logsdail, D. H., 234, 235, _277_
Longuet-Higgins, H. C., 81, _139_
Lonsdale, K., 20(66,67), _131_
Losty, H. H. W., 147, 149, 151, 147, 226, 227, 232, 243(141), _272_, _277_, _278_
Loubser, J. H. N., 8(23), 24-28(23,97,106), 30(23), 71(221), 110(221,283), _129_, _132_, _133_, _138_, _142_
Loudon, R., 51, _136_
Lungagnani, V., 184, 255, _275_
Lyons, G. A., 228(113), _276_

Magnus, Albertus, 111(288), _142_
Maiden, A. J., 52(163), 53(163, 164), _136_
Makarchenko, V. G., 163(51), _274_
Male, J. C., 55(182), 107, _136_, _141_
Mallick, P. K., _274_
Malmstron, C., 231, _277_
Mani, A., 52(168), 71(220), _136_, _137_
Mantell, C. L., 148, 206, _272_
Maradudin, A. A., 60(195), 62(195), 66(195), _137_
Marshall, P., 197, 203, 204, 209, 212(92), _276_
Martens, H. E., 231, 232, _277_
Martin, A. E., 9(36,37), 16(36), 17(36), 51(37), 52(36,37), 57(36), _130_
Martin, W. H., 248(144), 251(145), _278_
Mason, I. B., 163, 185, 226, 232(123), 233, _273_, _277_
Massier, P. E., 192, _275_
Matthews, I. G., 71(219), 88(219), 90(219), 91, 121(219), _138_
Matthews, R. B., 160, 243, 254, _273_
Maurer, R. J., 17(55), _131_
McClintock, F. E., 168, 176, _274_
McDonald, R. S., 24(95), _132_
McQuillan, A. K., 52(175), 53(175), _136_

McSkimmin, H. J., 67(214),
  69(214), 138
Medvedev, V. N., 71(218), 138
Melton, C. E., 5, 129
Messmer, R. P., 27-29(101, 108),
  34(101), 82(249), 133, 140
Meyer, R. A., 156, 164, 270(155,
  156), 273, 279
Miethe, A., 111(291), 142
Miller, W. H., 9, 130
Mingay, D. W., 4(18), 129
Mises, R. von, 161, 273
Mitchell, E. W. J., 33(124),
  34(124), 40(137), 43(124),
  71(219,223), 88(219), 90(219),
  91, 121(219), 134, 135, 138
Mitra, S. S., 52(173,174),
  53(173,174), 136
Miyazono, S., 241, 244, 278
Mobsby, C. D., 77(235), 113(235),
  115(235), 118(235), 119(235),
  139
Moore, E. B., 28(108), 133
Moore, M., 20(71), 131
Morgan, J. R., 178(74), 275
Morris, F., 155(30), 163(30),
  169(30), 170(30), 174(30),
  176(30), 244(30), 273
Morris, J. B., 229, 276
Mostoller, M., 60(198),
  61(198), 137
Mrozowski, S., 148, 232, 272
Murrell, S. A. F., 166, 167, 274
Mykolajewycz, R., 19(57,59), 131

Nabarro, F. R. N., 16(50),
  17(50), 57(50), 130
Nachal'naya, T. A., 27(98),
  110(283), 132, 142
Nadeau, J. S., 160, 273
Nadolinnyi, V. A., 110(283), 142
Nahum, J., 16(51), 17(51),
  110(285), 130, 142
Namjoshi, K. V., 52(174),
  53(174), 136
Narayanan, P. S., 52(170), 136
Nawi. O., 65(200), 67, 137
Nayar, P. G. N., 52(172),
  71(216), 136, 138
Nedvetskii, D. S., 82(246),
  90(258), 92(258), 140

Nepsha, V. I., 24(90), 30(90),
  108(90), 132
Nettley, P. T., 248, 251(145),
  278
Neuber, H., 222(104), 276
Newman, R. C., 24(93), 132
Newton, I., 111(289), 142
Nezbeda, C. W., 152, 272
Nightingale, R. E., 231, 277
Nikitin, A. V., 21(77), 22(77),
  131
Nilakantan, P., 20(65), 131
Norris, C. A., 71(219), 76(229),
  82(229), 83(229), 88(219),
  90(219), 91(219,229), 121(219),
  138, 139
Novikova, S. I., 67(213), 138
Novobratskaya, I. F., 244(142),
  278
Nuttall, R., 111(286), 142

Obert, L., 168, 274
O'Brien, M. C. M., 70, 81(241),
  138, 139
Ohtani, M., 205(99), 276
Okada, J., 153(26), 157(26), 273
Oku, T., 155, 158, 160(40), 273
Öpik, U., 27(102), 28, 81, 133,
  139
Orchard, J. S., 149, 157, 226,
  227, 243, 272
Orlov, Yu. L., 19(58), 131
Osantowski, J., 57(187), 137
Osten, von der, W., 60(199), 137
Owen, J., 110(284), 142

Parsons, B. J., 54(180), 136
Paterson, M. S., 179, 275
Pears, C. D., 148, 235, 272, 277
Peckham, G., 50(155), 135
Pekal'n, T., ,., 163(51), 274
Penchina, C. M., 77(234), 139
Petch, N. J., 163, 274
Peter, F., 9, 111(35), 130
Peterson, R. E., 222(107), 276
Phaal, C., 13(43), 20(43),
  21(43), 50(153), 130, 135
Philip, H. R., 57(186,188), 137
Pimkin, M. Ya., 205(98), 276
Piseri, L., 6(20), 50(20),
  51(20), 129

Platonov, P. A., 244(142), 278
Podol'skikh, L. D., 29(110), 133
Podzyarei, G. A., 27(98), 28(105),
    110(283), 132, 133, 142
Poferl, D. J., 30(115), 133
Pohl, R. W., 9, 16(33), 17(33),
    130
Powell, W. B., 192, 275
Price, M. S. T., 226, 276
Price, R. J., 174, 183, 245(143),
    274, 278
Priddle, E. K., 197, 203, 204,
    209(92), 212(92), 276
Pringsheim, P., 102(270), 141
Proctor, D. G., 153(25), 272
Pryce, M. H. L., 27(102), 28,
    37(134), 60(196), 81, 133,
    134, 136, 139

Raal, F. A., 5(4), 6, 8(23),
    12(4), 14(4), 17(56), 24-
    27)23), 30(23), 35(4), 50(4),
    113(19), 114(4), 128, 129,
    131
Rainey, P. H., 20(28b), 130
Ralph, J. E., 117(293), 142
Raman, C. V., 20(65), 52(166),
    131, 136
Ramanathan, K. G., 9, 10, 130
Ramaswamy, C., 52(169), 136
Ramdas, A. K., 50-52(158),
    54(158), 136
Rappeneau, J., 147, 272
Rauch, C. J., 33(124), 34(124),
    43(124), 134
Redfield, A. G., 17(53), 130
Reich, A. R., 207, 276
Reinkober, O., 9, 116(29), 130
Reiswig, R. D., 156(34), 239(34),
    273
Renk, K. F., 51(161), 136
Reynolds, W. N., 147, 148(7),
    151, 163(7,22), 232(122),
    235, 272, 277
Rimstidt, J. D., 36(132), 134
Ritter, J. T., 82(245), 140
Roberts, R. A., 57(189,190),
    58(189), 137
Robertson, R., 9, 16(36), 17(36),
    51(37), 52(36,37), 57(36),
    130

Robinson, E., 205, 276
Rocco, G. G., 5(9), 129
Röder, N., 60(199), 137
Roessler, D. M., 57(190), 137
Romestain, R., 71(212), 138
Ross, A. M., 232(125), 277
Roundthwaite, C., 228(113), 276
Rowe, G. W., 233, 277
Ruffino, G., 37(135), 134
Runciman, W. A., 5(11), 49(11),
    50(11), 71(217), 78(217),
    90(217,224), 129, 138
Ruvald, J., 54(179), 136
Rykov, A. N., 9(27), 49(27),
    129
Ryneveld, van, W. P., 28(106),
    133

Sack, R. A., 81, 139
Saito, M., 263, 278
Salkovitz, E. I., 165, 274
Salotti, C. A., 5(10), 129
Samoilovich, M. I., 21(77),
    22(77), 29(110), 131, 133
Samsonenko, N. D., 24(88), 25(88),
    110(283), 132, 142
Sato, S., 205, 232, 241, 244, 276
    277, 278
Savaria, L. R., 57(191), 137
Schastnev, P. V., 110(283), 142
Schonland, D. S., 28(107), 133
Seal, M., 5(12), 24(12), 129
Seldin, E. J., 150, 153(25), 272
Sellschop, J. P. F., 4(3,18),
    5(3), 49(3), 128, 129
Semenov, A. G., 110(283), 142
Shapiro, O. Z., 110(283), 142
Sharma, J., 36(132), 134
Shcherbakova, M. Ya., 110(283),
    142
Sherman, W. F., 35(128), 36(128,
    131), 47(128,131), 114(131),
    134
Shevchenko, S. A., 17(52),
    57(52), 130
Shul'man, L. A., 27(98), 28(105),
    110(283), 132, 133, 142
Sibley, W. A., 60(198), 137
Simeral, W. G., 9(25), 10(25),
    14(25), 15(25), 50(25),
    112(25), 129

Simmons, J. H. W., 151, 232(122), 248(144), 272, 277, 278
Simon, F. E., 23(84), 132
Sines, G., 168, 174, 274
Skinner, B. J., 67(213), 138
Slack, G. A., 23, 132
Slagle, O. D., 154, 273
Smakula, A., 19(57,59), 131
Smirnov, L. S., 107(279), 141
Smith, H., 20(66,67), 131
Smith, H. M. J., 6(22), 23(22), 34(22), 47(22), 129
Smith, J. E., 150(13), 170(13), 272
Smith, M. C., 148, 157, 235, 237, 238, 272, 277
Smith, R. A., 57(185), 137
Smith, S. D., 6(21), 23(21), 24(94), 31(118), 32, 33(138), 34(21), 35(118), 37(21,118, 135), 47(21), 50(21), 51, 54(21), 64(21, 112(297), 114, 129, 132, 134, 135, 135, 136, 143
Smith, W. V., 27(96), 28(96), 132
Smitnov, L. S., 30(115), 133
Snowdon, R. A., 228(113), 276
Sobolev, E. V., 9, 11, 13(42), 18(41), 20, 21, 24(88), 25(88), 71(222), 99(265), 110(283), 118(295), 130, 132, 138, 141, 142
Solin, S. A., 50-52(158), 54(158), 77(236), 136, 139
Sorokin, L. A., 17(52), 57(52), 130
Sorokin, P. P., 27(96), 28(96), 132
Soule, D. E., 152, 272
Spence, G. B., 153(25), 272
Squires, R. L., 228, 276
Srawley, J. E., 196, 275
Stachurski, Z. H., 197, 203(93), 235(93), 276
Stanley, R., 36(132), 134
Stenman, F., 52(176), 136
Stevens, R., 198, 203, 276
Stoicheff, B. P., 52(175), 53(175), 136

Stoneham, A. M., 27(100), 80(237,238), 81(239), 82(248, 249), 93(260), 94, 96(260), 132, 139, 140, 141
Stranningan, J. S., 202(95), 276
Straumanis, M. E., 5, 129
Strong, H. M., 24(89), 25(89), 27(89), 30(113), 36(113), 47(113), 108(113), 132, 133
Sturge, M. D., 37(134), 134
Summers, C. J., 33(138), 135
Summersgill, I., 15(47), 17(47), 19(47), 72(47), 88(47), 123(47), 130
Sutherland, G. B. B. M., 9, 10(25,40), 14(25), 15(25), 50(25,40), 112(25), 129, 130
Sutton, A. L., 232(121), 277
Sutton, J. R., 1(1a-f), 128
Swift, P., 36(132), 134
Symons, M. C. R., 29(109), 133

Taft, E. A., 57(186,188), 137
Takagi, M., 21(75), 131
Tappin, G., 194, 203, 275
Tattersall, H. G., 194, 203, 275
Taylor, R., 155(30), 163(30), 169(30), 170(30), 174(30), 176(30), 244(30), 273
Taylor, W., 31(118), 32, 34, 35(118), 37(118,135), 114, 134
Thewlis, J., 67(213), 138
Thomas, D. G., 106(273,276), 141
Thomas, J. M., 36(132), 134
Thomaz, M. F., 99(267), 120(267), 141
Thomaz, M. F., 99(267), 120(267), 141
Thrower, P. A., 148(7), 163(7), 272
Titova, V. M., 21(76), 22(76), 131
Tolansky, S., 30(112), 133
Tsuzuku, T., 263, 278
Tubino, R., 6(20), 50(20), 51(20), 129
Tucker, M. O., 224, 276
Tuft, 24(89), 25(89), 27(89), 132

Turk, L. A., 22(79), 23(79), _132_
Turnbull, A., 263(153), _278_

Ubbelohde, A. R., 258(148), _278_
Udovskii, A. L., 195, 203(86),
    235(86), 236(86), _275_
Uhlenbeck, G. E., 85(252), _140_
Umeno, M., 36(133), 58(133), _134_

Valachovic, R. S., 195, 196, 203,
    204(89), _275_
Valkenburg, van, A., 5, _129_
Vella-Coleiro, G., 33(138), _135_
Verma, G. S., 23, _132_
Vermeulen, L. A., 16(50), 17(50),
    57(50), 76(227,228,271),
    82(247), 103, 105, _130_, _139_,
    _140_, _141_
Virgil'ev, Yu. S., 163, 244(142),
    _274_, _278_
Vitovec, F. H., 195, 197,
    203(93), 235(93), _275_, _276_
Volta, G., 230, _277_
Vorozheikin, K. F., 21(77),
    22(77), _131_

Walker, J., 76(227,229), 77(233),
    78(233), 115(233), 117(296),
    119(296), _139_, _143_
Walker, W. C., 57(187,189,190),
    58(189), _137_
Walsh, J. B., 168, 177, _274_
Walsh, P. S., 47(151), 108(281),
    110(281), _135_, _142_
Walter, B., 71, 111(215), _138_
Warren, J. L., 50(154), 51(154),
    55(154), _135_
Watkins, G. D., 27-29(101,108),
    34(101), 68(208), 82(249),
    110(283), _133_, _138_, _140_, _142_
Wedepohl, P. T., 31(120), _134_,
    _135_
Wedlake, R., 67(203), 90(203),
    116(203), _137_
Weeks, J. C., 259(149), 260(149),
    _278_
Weertman, J., 157(36), 239(36,
    135), _273_, _278_

Wehner, R., 6(21), 23(21),
    34(21), 37(21), 47(21),
    50(21,157), 54(21), 67(21),
    _129_, _136_
Weibull, W. J., 167, 181, _274_
Wells, D., 153, _273_
Weng, T., 153(25), _272_
Weng, Tu-Lung, _274_
Wentorf, R. H., 30(113), 36(113),
    47(113), 108(113), _133_
Wenzel, R. G., 50(154), 51(154),
    55(154), _135_
Wessel, E. T., 196, _275_
Whetton, N. R., 57, _137_
Whiffen, D. H., 27(99), _132_
Whippey, P. W., 24(93), _132_
Wiech, G., 36(133), 58(133), _134_
Wight, D. R., 31(116), 43-45(116),
    47(116,147), 55(116), 56(116),
    64(147), 77(235), 113(235),
    115(235,292), 118(235),
    119(235), 122(292), 123(116,
    292), _134_, _135_, _139_, _142_
Wild, A. M. A., 110(283), _142_
Wild, R. K., 93(259), _140_
Wilkins, B. J. S., 160, 206(44),
    207, 218, _273_, _276_
Williams, A. W. S., 31(121),
    32(121), 34, 37(121),
    45(121), 46(121), 47(146,
    149), _134_, _135_
Williams, F. E., 106(274), _141_
Williamson, G. K., 151, _272_
Willis, H. A., 10, 50(40), _130_
Winer, S. A. A., 108(282),
    110(282), _142_
Wobschall, D. C., 153(26),
    157(26), _273_
Wolfe, R., 31(119), _134_
Wood, R. F., 60(198), 61(198),
    _137_
Woods, W. K., 243(140), _278_
Wooley, R. L., 184, _275_
Wooster, W. A., 20(69), _131_
Wright, A. C. J., 71(221),
    110(221,283), _138_, _142_
Wright, C. E., 13(44), 20-22(44,
    70), _130_, _131_
Wu, C. H., 54(179), _136_

Yahr, G. T., 150(13), 170(13), 195, 196, 203, 204(89), 272, 275

Yarnell, J. L, 50(154), 51(154), 55(154), 135

Yu'eva, O. P., 118(295), 142

Zacks, E., 107(277), 141

Zaitsev, A. L., 244(142), 278

Zaitsev, G. G., 205(98), 276

Zaritskii, I. M., 28(105), 133

Zerbi, G., 6(20), 50(20), 51(20), 129

Ziman, J. M., 23(80,84), 132

Zimmer, J. E., 270(155,156), 279

Zubkov, V. M., 9(27), 49(27), 129

Zukas, E. G., 156(34), 157(35, 36), 239(34,35,36,135), 273, 278

Zyl, van, C., 2(2), 128

SUBJECT INDEX

Absorption spectrum, 6, 57, 61, 63, 74, 81, 83, 104, 124
Acoustic emission, 153
Age of diamonds, 2
Aluminum in diamonds, 6, 46
Ammonia inversion line, 85
Annealing, 237

Bend strength, 186, 224, 247
Biaxial strength, 170
Binder in graphite, 147
Blue diamond, 36, 47
Blue luminescence, 107
Boron in diamond, 46
Broad bands, 113
Bromine intercalation, 259

C_{44}, 152
Cathodoluminescence, 42, 47, 101, 107, 126
Carbon
  fibers, 150
  glassy, 148, 170
Compressive strength, 158, 169, 244
Cracks
  Griffith crack theory, 161, 163, 166, 181, 225, 230, 248, 266
  growth of, 146, 228
  initiation of, 146, 232, 237
  theory of failure, 163
Creep, 152, 156, 237, 239

Debye cut-off, 65
Deformation, 150
Density of graphite, 148, 225

Diamonds
  age of, 2
  aluminum in, 6, 46
  blue, 36, 47
  boron in, 46
  coat, 24
  gold in, 5
  green, 5, 22
  impurities in, 2, 4
  inclusions in, 4
  Type Ia, 8, 93
  Type IIa, 48
  Type Ib, 24
  Type IIb, 30, 40, 46
Dislocations, 150, 163, 227, 243, 248
Dynamic fatigue, 160

Electrical resistivity, 159
Electron
  irradiation, 97
  microscopy, 19, 156, 198, 232
  paramagnetic resonance, 110
Emission spectrum, 84
Endurance level, 213

Failure mechanisms, 160
Fatigue, 204
  dynamic, 160
Fracture, 165
  mechanics, 193, 266
  micro-, 157
  toughness, 202
  work of, 194, 202, 235, 240
Friction, internal, 151

Gilsocarbon, 156, 159, 245, 255
Glassy carbon, 148, 170

293

Gold in diamond, 5
Grain size, 225, 230, 267
Graphite
    binder in, 147
    density of, 148, 225
    isotropic, 155, 171, 198
    pyrolytic, 152, 235
    structure of, 146
Green
    diamonds, 5, 22
    luminescence, 107
Griffith crack theory, 161, 163,
    166, 181, 225, 230, 248,
    266

Heat treatment, 227
Homologous stress, 205
Huang-Rhys factor, 60
Hydrostatic pressure, 178
Hysterisis, mechanical, 151

Impact testing, 240
Impregnation, 226
Impurities in diamond, 2, 4
Inclusions in diamond, 4
Infrared spectra, 8, 25, 30
Intercalation, 258
Intermediate type diamonds, 49
Internal friction, 151
Inversion splitting, 84
Irradiation
    and mechanical properties,
        169, 174, 177, 207, 209,
        219, 241, 243, 269
    and UV absorption in diamond,
        75
    electron, 97
    neutron, 153, 158, 219, 243,
        269

Jahn-Teller effect, 27, 68, 80,
    90

Kirchoff's law, 52
Kramers-Kronig analysis, 57

Luminescence, 2, 19, 40, 51, 55,
    61, 63, 71, 76, 83, 86,
    91, 96, 99
    blue, 107
    cathodo-, 42, 47, 101, 107,
        126
    green, 107
    photo-, 84, 111
    polarized, 71
    thermo-, 110
    tribo-, 110

Magnetic properties of diamond,
    19
Microcracks, 148, 225, 242, 265
Microfracture, 157
Microscope
    electron (see Electron
        microscopy)
    optical, 154

Neutron
    irradiation, 153, 158, 219,
        243, 269
    scattering, 50
Nitrogen
    atomic form, 18, 72, 82
    in diamond, 2, 11, 18, 48,
        71, 92, 99, 107
Notch sensitivity, 218, 266

Opaque diamond, 36
Oxidation, 228, 256

Paramagnetism, 30
Petch formula, 163
Petroleum coke, 154, 155, 163,
        226, 233
Photochromic effect, 103
Photoconduction, 16, 36, 57, 103
Photoluminescence, 84, 111
Polarized luminescence, 71
Pressure, hydrostatic, 178
Prestress, 157
p-Type semiconduction, 32
Pyrolytic graphite, 152, 235

Radiation damage in diamond,
    12, 19, 75, 82, 88, 96,
    102, 115, 122
Raman spectroscopy, 18, 44,
    52
Raman Grüneisen parameter, 53
Refractive index, 111
Resistivity, electrical, 159
Rotary bending, 205

Santa Maria coke, 229
Self-diffusion, 238
Semiconduction, p-type, 32
Sensitivity, notch, 218, 266
Shear, 151, 153, 170
Slip systems, 161
Space Group, 52
Spectra
    effect of temperature on,
        14, 67
    infrared, 8, 25, 30
    UV, 14, 26, 40, 105
Spectrum
    absorption, 6, 57, 61, 63,
        74, 81, 83, 104, 124
    emission, 84
Stark effect, 71
Strain
    broadening, 93
    rate, 231, 234, 237, 239
Strength, 157, 202, 263
    bend, 186, 224, 247
    biaxial, 170
    compressive, 158, 169, 244
    effect of temperature on,
        231, 236
    tensile, 158, 169, 186,
        224, 244
    triaxial, 176
    uniaxial, 169
    vacuum, effect of on, 233
Stress
    effect on spectra, 69, 78
    homologous, 205
    prestress, 157
    thermal, 192
    uniform, 165

Surface energy, 161, 234
Synthetic diamond, 108

Temperature
    spectra, effect on, 14, 67
    strength, effect on, 231, 236
Tensile strength, 158, 169, 186,
        224, 244
Thermal
    conductivity, 22, 252
    shock, 240
    stress, 192
Thermoluminescence, 110
Torsion, 205
Triaxial strength, 176
Triboluminescence, 110
Tubular specimens, 170
Type Ia diamond, 8, 93
Type IIa diamond, 48
Type Ib diamond, 24
Type IIb diamond, 30, 40, 46

Ultraviolet spectra, 14, 26, 40,
    105
Uniaxial strength, 169
Uniform stress, 165

Vacuum, effect on strength, 233
Volume charge, irradiation
    induced, 251
Volume-strength relationship,
    184
Von Mises criterion, 161

Wannier exciton, 42
Water vapor, 234
Weibull theory, 174, 181
Work of fracture, 194, 202,
    235, 240

Young's modulus, 149, 157, 202,
    225, 263